Electrical Machines

Electrical Installat[illegible]vanced Course

E. G. Stocks and J. N. Hooper

Edited by Chris Cox

THOMSON LEARNING

Australia · Canada · Mexico · Singapore · Spain · United Kingdom · United States

THOMSON LEARNING™

Electrical Machines

For more information, contact Thomson Learning, Berkshire House, 168–173 High Holborn, London, WC1V 7AA or visit us on the World Wide Web at: http://www.thomsonlearning.co.uk

---

British Library Cataloguing-in-Publication Data
A catalogue record for this book is available from the British Library

---

ISBN 1-86152-727-6

First published 2001 by Thomson Learning

Printed in Croatia by Zrinski d.d.

About this book

"Electrical Machines" is one of a series of books published by Thomson Learning related to Electrical Installation Work. The series may be used to form part of a recognised course, particularly City and Guilds 2360 Electrical Installation, or individual books can be used to update knowledge within particular subject areas. A complete list of titles in the series is given below.

Electrical Installation Series

Foundation Course

Starting Work
Procedures
Basic Science and Electronics

Supplementary title:
Practical Requirements and Exercises

Intermediate Course

The Importance of Quality
Stage 1 Design
Intermediate Science and Theory

Supplementary title:
Practical Tasks

Advanced Course

Advanced Science
Stage 2 Design
Electrical Machines
Lighting Systems
Supplying Installations

Acknowledgements

Every effort has been made to trace all copyright holders but if any have been inadvertently overlooked, the publishers will be pleased to make the necessary arrangements at the first opportunity.

Study guide

This studybook has been written to enable you to study either in a classroom or in an open or distance learning situation. To ensure that you gain the maximum benefit from the material you will find prompts all the way through that are designed to keep you involved with the subject. If you are studying by yourself the following points may help you.

- Work out when, and for how long, you can study each week. Complete the table below and from this produce a programme so that you will know approximately when you should complete each chapter. Your tutor may be able to help you with this. It may be necessary to reassess this timetable from time to time according to your situation.
- Try not to take on too much studying at a time. Limit yourself to between 1 hour and 2 hours and finish with a Try this or the short answer questions (SAQ) at the end of the chapter. When you resume your study go over this same piece of work before you start a new topic.
- You will find answers to the questions at the back of the book but before you look at the answers check that you have read and understood the question and written the answer you intended.
- An end test is included so that you can assess your progress.
- Try this activities are included and you may need to ask colleagues at work or your tutor at college questions about practical aspects of the subject. These are all important and will aid your understanding of the subject.
- It will be helpful to have available for reference a current copy of BS 7671:1992. At the time of writing this incorporates Amendment No.1, 1994 (AMD8536), Amendment No. 2, 1997 (AMD 9781) and Amendment No. 3 (AMD 10983) 2000.
- Your safety is of paramount importance. You are expected to adhere at all times to current regulations, recommendations and guidelines for health and safety.

Study times					
	a.m. from	to	p.m. from	to	Total
Monday					
Tuesday					
Wednesday					
Thursday					
Friday					
Saturday					
Sunday					

Programme	Date to be achieved by
Chapter 1	
Chapter 2	
Chapter 3	
Chapter 4	
Chapter 5	
Chapter 6	
Chapter 7	
Chapter 8	
End test	

Contents

1

Machines – Basic Theory

At the beginning of all the other chapters in this book you will be asked to complete a revision exercise based on the previous chapter.

To start you off we will use this opportunity to remember some facts that you should be aware of.

The unit of magnetic flux is the ____________, and the unit of magnetic __________ density is the ____________.

Fleming's L.H. Rule can be used to determine the________ in which a current-carrying ________________ tends to __________ when it is placed in a ________ ______.

A generator converts __________ __________ into ______________ ______________ and a motor converts _____________ ______________ into ______________ ______________.

The ____________ effect, or torque, produced by a.c. induction motors, either __________ phase or ________ phase, depends upon the __________ of __________ magnetic fields.

Direct-on-line ____________ are used for __________ the majority of small three-phase ___________ __________ ________ motors.

All motors having a rating exceeding ________ must have control equipment with __________ __________.

No-volt protection is provided to prevent ____________ ____________ of machinery after a supply __________.

On completion of this chapter you should be able to:

- identify the basic rules of electromagnetism
- describe a magnetic circuit
- calculate the effectiveness of a magnetic circuit
- describe the relationship between magnetism, current flow and mechanical movement
- describe the effects of self inductance

Basic theory

Most people are first introduced to d.c. motors when they play with toy trains and racing cars. These are often supplied directly from a battery or a small transformer rectifier unit. The speed control for the toys is usually through a variable resistor that adjusts the voltage to the motor. To reverse the direction of travel a switch is incorporated that changes the supply polarity to the motor and reverses its direction of rotation.

Figure 1.1

As these toys are driven by d.c. motors it is easy to run away with the idea that all d.c. motors can have their speed controlled by varying the supply voltage, and their direction of rotation reversed by changing the supply polarity. Unfortunately the small permanent magnet motor that is used in these toys is the only type that reacts in this way.

Revision of magnetic theory

Permanent magnets

As d.c. machines rely on magnetism to make them work, before continuing with their actual construction it may be useful to revise the basic magnetic principles.

Figure 1.2 shows the magnetic field around a permanent magnet.

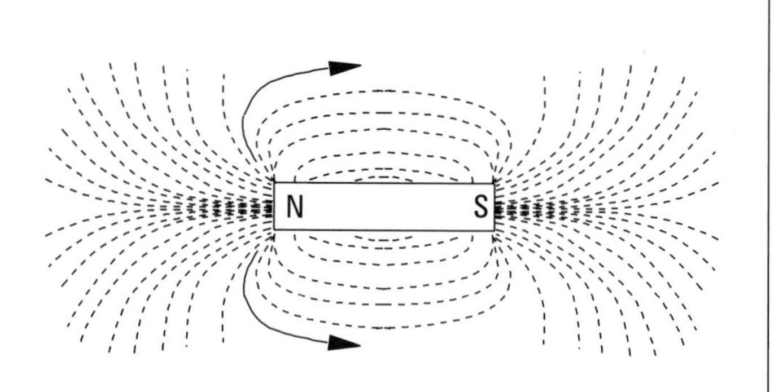

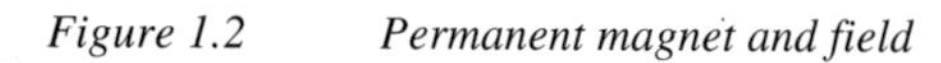

Figure 1.2 *Permanent magnet and field*

The flux lines have properties that must always be considered, these include:

- all magnets have both North and South poles
- the lines of flux are assumed to have a direction of South to North through the core of the magnet but North to South outside the magnetic poles
- each line of flux is complete, that is to say it is a complete circuit, part of it being inside the magnetic material the rest being through the air
- each line of flux is separate
- the lines of force from a magnet cannot be destroyed but they can be concentrated or distorted

There are other factors but the above all relate to the working of machines.

When two permanent magnets are fixed so that a north pole faces a south pole, a resultant field distribution takes place.

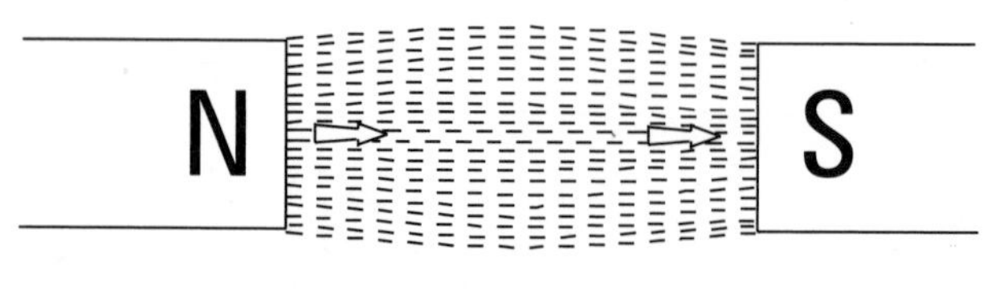

Figure 1.3 *The combined magnetic field distribution related to two magnets*

Electromagnets

When an electrical conductor carries a direct current there is always a magnetic field related to it. A single conductor has a field distribution as shown in Figure 1.4.

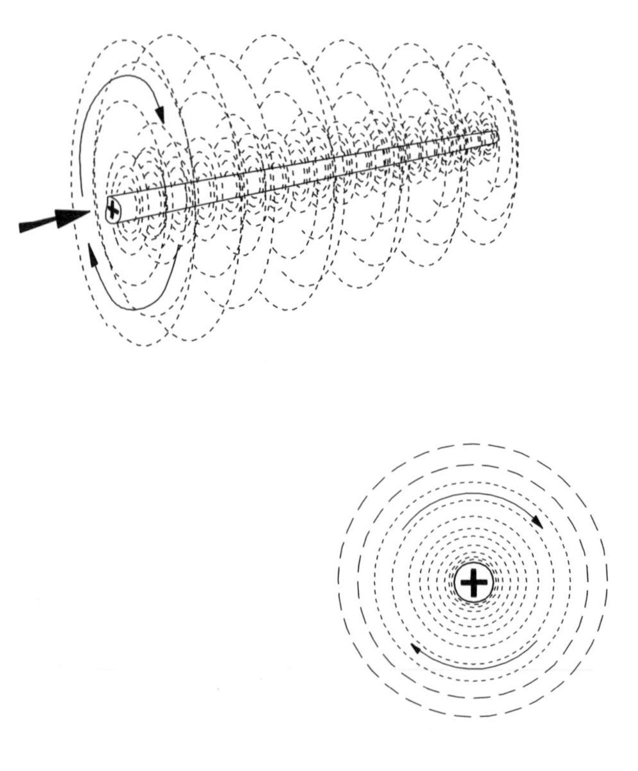

Figure 1.4 *The magnetic field distribution around a single current carrying conductor*

Although for simplicity only a sample of fields are shown, these are, in practice, continuous throughout the conductor's length. The magnetic field is strongest close to the conductor and gets weaker further away. If a current carrying conductor is wound as a single coil the resultant magnetic field is as shown in Figure 1.5.

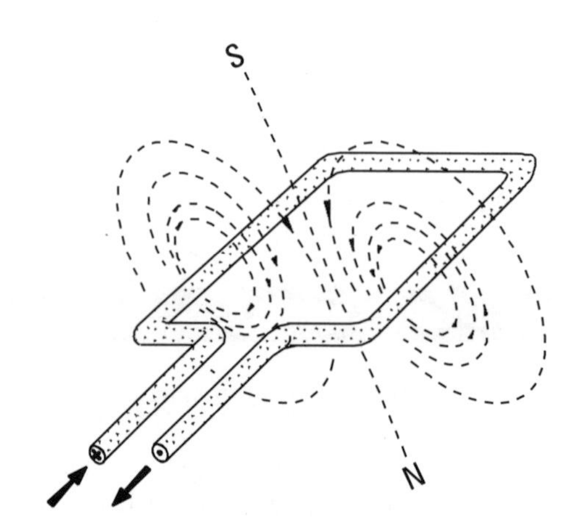

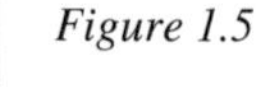

Figure 1.5 *The magnetic field associated with a single turn of wire carrying a direct current*

When looking at the field at this point the current is in one direction in half of the turn and it is going in the other direction for the other half. This gives the effect of a centre north soulh line. If now a number of turns are shown, the magnetic field distribution is the result of each single turn added together as shown in Figure 1.6.

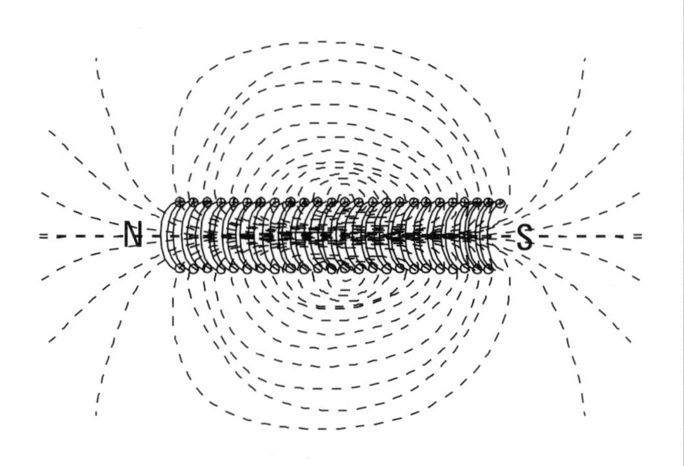

Figure 1.6 *The magnetic field distribution related to a coil of wire carrying a direct current*

The resultant field distribution related to this is not unlike that of the permanent magnet but is not so concentrated. However if a bar of magnetic conductive material is placed inside the coil this changes.

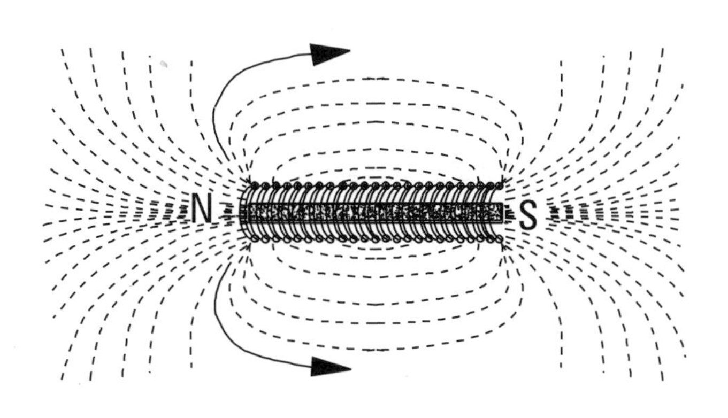

Figure 1.7 *The magnetic field distribution related to a current carrying coil an iron core*

The iron core has the effect of concentrating the field into definite poles at each end of the coil. The properties of the magnetic field are the same as those for the permanent magnet.

Remember

Magnetic conductive materials used for cores in this way would have a high relative permeability.

If you require further information on this subject refer to the studybook "Advanced Science" in this same series.

Basic motor theory

In order to create movement the current carrying conductor must now be placed in a magnetic field. Using the two facing poles as shown in Figure 1.3 the magnetic fields combine and produce a force on the conductor (Figure 1.8).

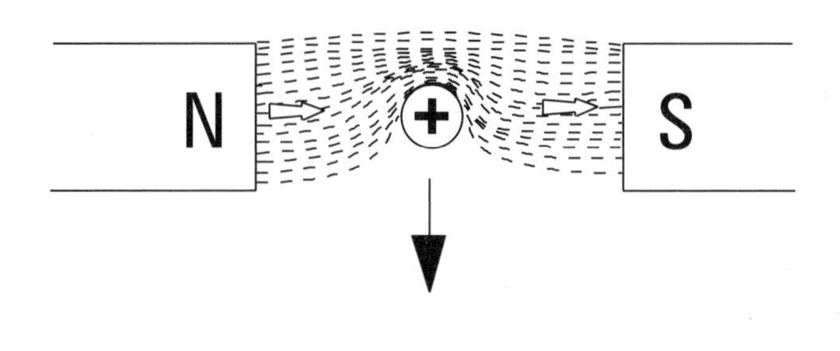

Figure 1.8 *The force produced on a current-carrying conductor in a magnetic field.*

If the direction of the current is reversed, the force on the conductor is also reversed, as Figure 1.9 shows.

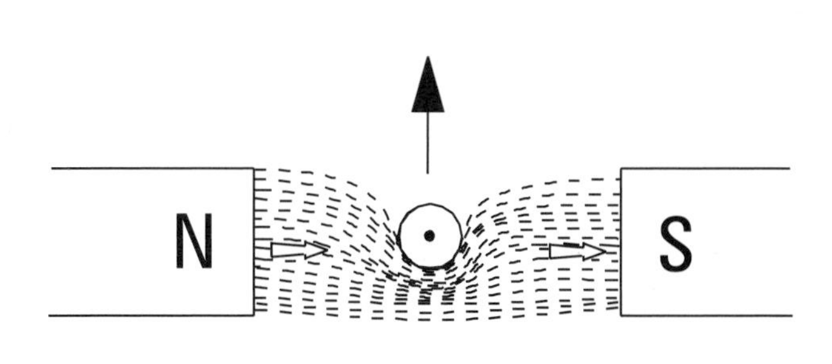

Figure 1.9 *Reversing the current reverses the force.*

In Figure 1.9 the North and South poles of the magnet were left as they were in Figure 1.8, but if these were reversed as well the result would be as shown in Figure 1.10.

Now it can be seen that the force on the conductor is the same as in Figure 1.8 even though everything has been reversed.

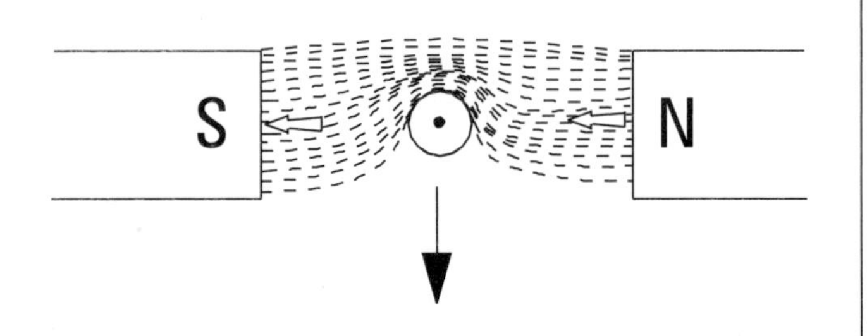

Figure 1.10 *Both the magnetic poles and polarity of the current are changed from that shown in Figure 1.8.*

The direction of rotation related to the fixed magnetic field can be determined using "Fleming's Left Hand Rule" as shown in Figure 1.11.

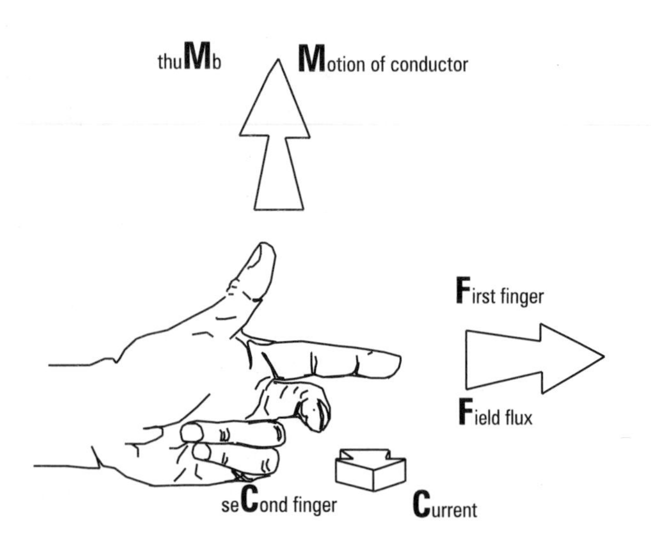

Figure 1.11 *Fleming's Left Hand Rule*

Remember

MOTORS drive on the **LEFT!**

The magnitude of the force produced on the conductor can be calculated from

$$F = B I L$$

where

F	=	force in Newtons (N)
B	=	strength of magnetic field in Teslas (T)
I	=	current in Amperes (A)
L	=	length of conductor affected by the magnetic field (m)

Example

A single conductor has a length of 0.1 m inside a magnetic field with a strength of 0.8 T. When a current of 5 A is passed through the conductor a force is produced of:

$$F = B I L$$
$$= 0.8 \times 5 \times 0.1$$
$$= 0.4 \text{ N}$$

Rotation

To create a rotating motion a coil of current carrying conductor has to be placed inside the magnetic field. To examine this a single turn of conductor is used in Figure 1.12. As in effect the single turn has current going away in one half but coming towards in the other, the two halves of the coil are forced in different directions.

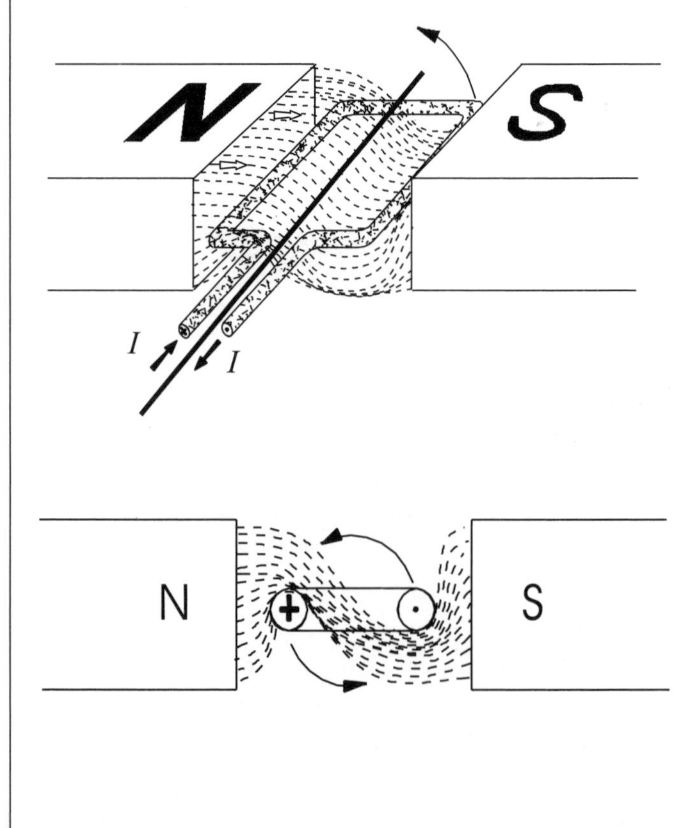

Figure 1.12 *The rotating force produced on a single coil in a magnetic field*

Assuming there is a shaft for the coil to rotate on, the coil would be forced to a position at 90° to the fixed magnetic field as shown in Figure 1.13. When a coil is rotated in this way it is said to produce a torque.

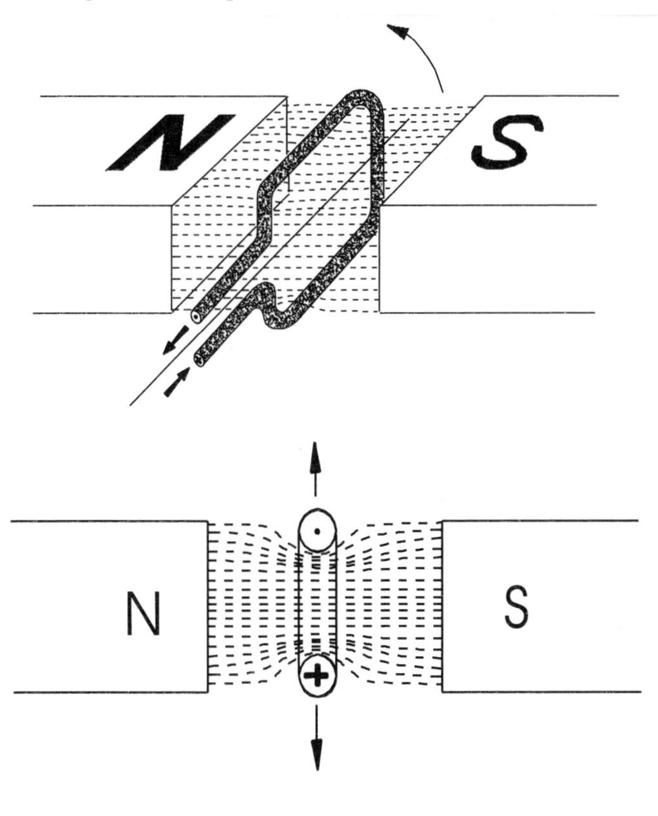

Figure 1.13 *The coil in Figure 1.12 has now rotated 90°.*

Commutation

To get past this point a second coil is used placed at 90° to the first. Current is switched off to the first coil and on to the second. The force on the second coil now rotates the assembly until it is at 90° to the fixed magnetic field. If the direction of rotation is to be maintained the current has to be switched off in the second coil and back on in the first but now in the reverse direction. This switching is carried out automatically with a rotating switch called a commutator. This arrangement is shown in Figure 1.14.

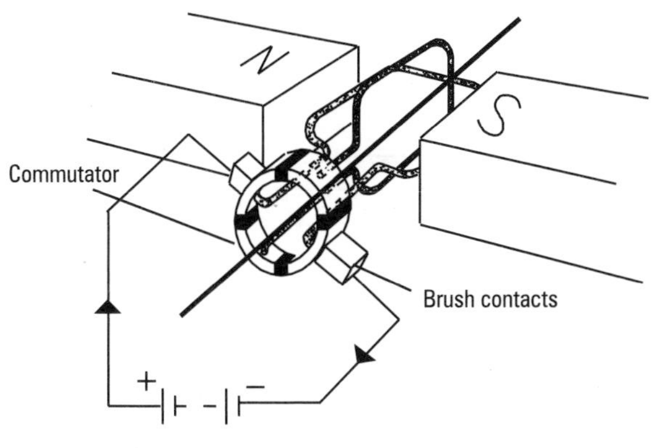

Figure 1.14 *A simple armature arrangement with commutator.*

Try this

If a conductor, with a total length of 2.5 m inside a magnetic field of strength of 0.62 T, has a current of 11.4 A flowing through it what is the force on the conductor?

Basic generator theory

Some of the basic theory related to motors also applies to generators. Where a motor uses the flow of current to produce mechanical movement, a generator uses the mechanical drive from some other source to move an electrical conductor through a magnetic flux and thus create a current flow.

When any electrical conductor is moved through a magnetic field an electromotive force (e.m.f.) is induced into it. Figure 1.15 shows a conductor being moved through the magnetic field between two poles.

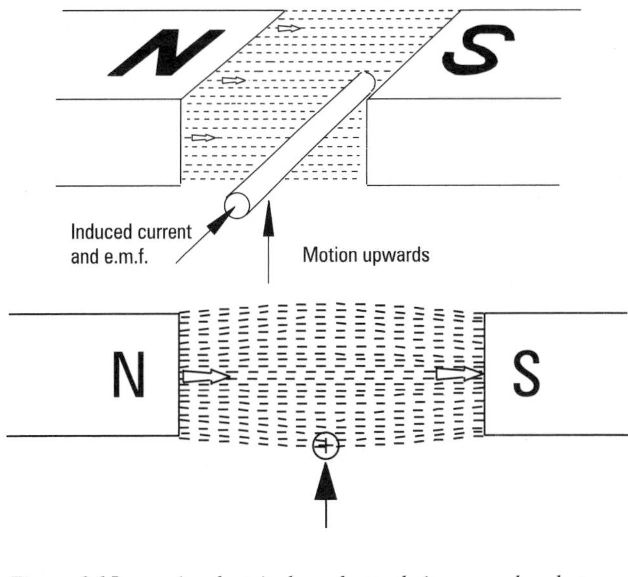

Figure 1.15 *An electrical conductor being moved up between two magnetic poles.*

This can be remembered by using Fleming's Right Hand Rule as shown in Figure 1.16.

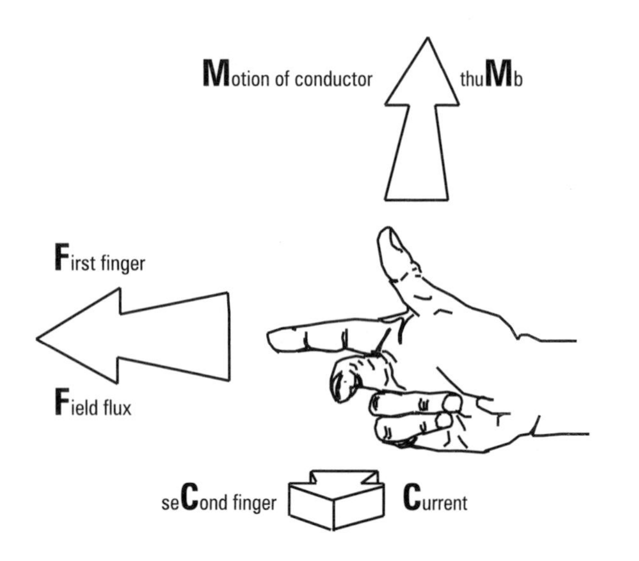

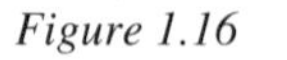
Figure 1.16 Fleming's Right Hand Rule

Generated e.m.f.

The instantaneous value of induced e.m.f. can be calculated from

$$e = Blv$$

where

e = the instantaneous value of e.m.f. in volts (V)
B = magnetic flux density in Tesla (T)
l = the length of conductor in the magnetic field in metres (m)
v = the velocity of the conductor through the field in metres per second (m/s)

Example

When a conductor 0.4 m long is moved at a velocity of 14 m/s through a magnetic flux which has a strength of 0.8 T it produces an e.m.f. of

$$\begin{aligned} e &= Blv \\ &= 0.4 \times 14 \times 0.8 \\ &= 4.48 \text{ V} \end{aligned}$$

To produce a continuous output the conductor must be moving in the same direction through the same magnetic field strength. Although this is impracticable to achieve, a compromise can be obtained using a coil. The conductor, when shaped into a loop, has at least twice the length inside the magnetic field. When it is rotated an e.m.f. is induced into it. As the direction of movement is not always going to be 90° to the magnetic field, the actual angle must be taken into account. The instantaneous induced e.m.f, when there is an angle involved, can be calculated from

$$e = Blv \sin\theta$$

where $\sin\theta$ is the angle between the direction of the magnetic flux and the direction of movement.

Try this

Calculate the e.m.f. induced into a single conductor 0.25 m long which is moved at velocity of 16 m/s at an angle of 45° when in a magnetic flux of 0.85 T.

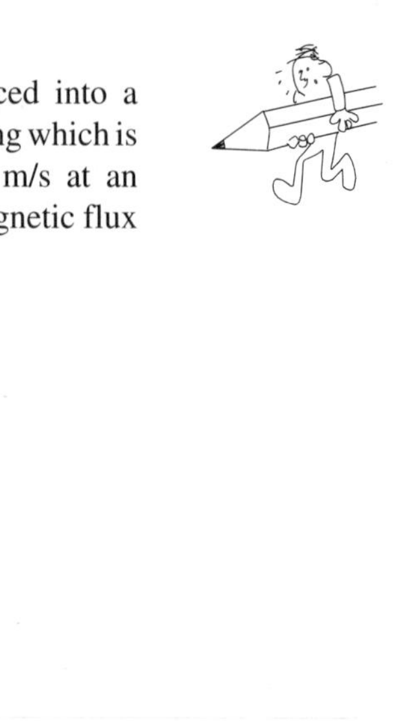

Figure 1.17 shows a loop of wire being rotated on a shaft and the generated e.m.f. is being measured from slip rings connected to either end of the conductor.

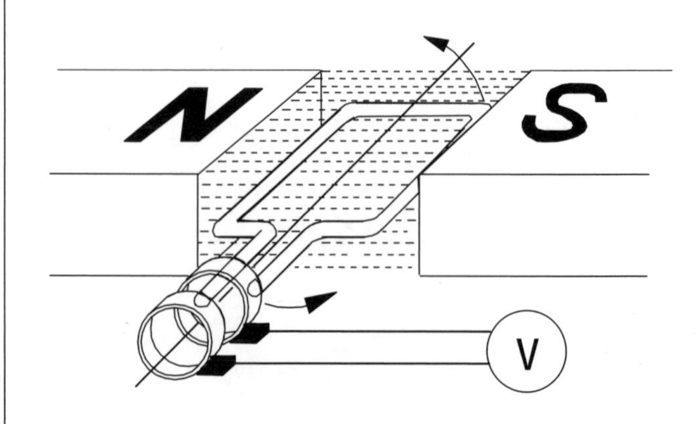

Figure 1.17 a.c. generator

It can be seen that the output of this arrangement, plotted in Figure 1.18, is an alternating one and not direct current.

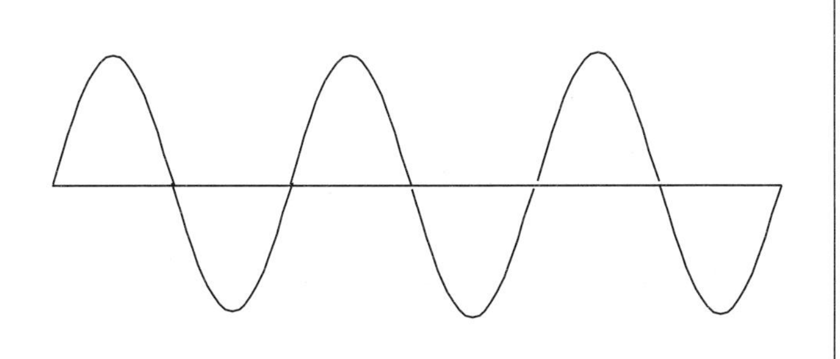

Figure 1.18 *a.c. waveform output from the coil in Figure 1.17.*

This is the basic theory of the a.c. generator (alternator).

To generate d.c. the lower section of the waveform has to be reversed using a changeover switch. This changeover switch is a commutator fitted in place of the slip rings used previously in Figure 1.17. To give a more constant output several sets of windings are fitted to the armature, each having their own segments on the commutator. Figure 1.19 shows a simple d.c. generator with two wire loop windings and Figure 1.20 shows the output from such an arrangement.

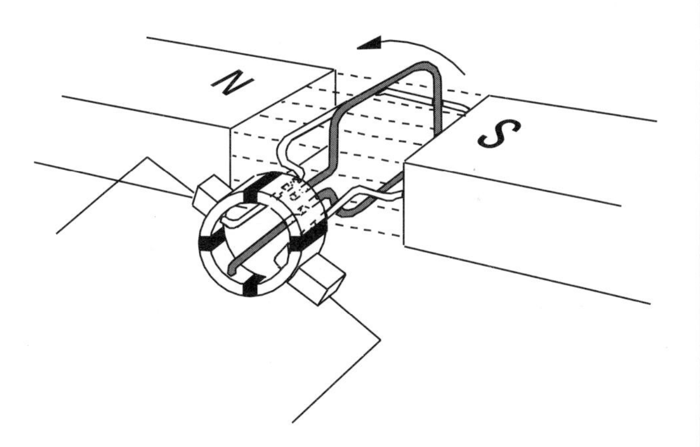

Figure 1.19 *d.c. generator*

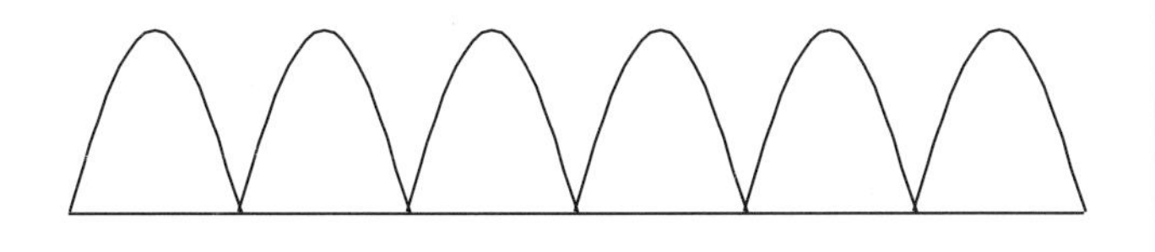

Figure 1.20 *Output from the generator in Figure 1.19.*

Motors and generators

It can now be seen that the actual construction of a motor and a generator is basically the same. It follows then that when a motor is powered from a supply and is running at full speed it will also be generating a voltage. As it can be determined from Lenz's Law, this induced e.m.f. is opposite in direction to the supply voltage. It is known as the "back e.m.f.". This generated voltage is always less than the supply due to the voltage drop in the armature windings. The voltage drop in these windings can be calculated using Ohm's Law.

Armature voltage drop
= armature current × armature resistance

$= I_a \times R_a$

It can also be shown that

Armature voltage drop
= supply voltage – back e.m.f.

$I_a R_a = U - E$

where

U = supply voltage
E = back e.m.f.
I_a = armature current
R_a = armature resistance

To calculate the supply voltage for a given motor the formula can be transposed so that

$U = E + I_a R_a$

Example

When connected to a 200 V supply a d.c. motor has a current of 20 A flowing through the armature which has a resistance of 0.3 ohms. Calculate the back e.m.f. generated in the motor.

First the formula must be transposed so that the back e.m.f. is the subject.

This can be carried out from $I_a R_a = U - E$.

so then $E = U - I_a R_a$

Filling in the details

$E = 200 - 20 \times 0.3$
$= 200 - 6$
$= 194$ V

Remember

Lenz's Law

The direction of an induced e.m.f. is always such that it sets up a current opposing the change of flux responsible for inducing that e.m.f.

Try this

1. A d.c. motor has an armature which has a voltage drop of 6 V when connected to a 250 V d.c. supply. Calculate the back e.m.f. of the motor.

2. An armature in a d.c. motor has a resistance of 0.2 ohms and creates a voltage drop of 6.5 V when the machine is connected to a 240 V d.c. supply. Calculate
   (a) the back e.m.f. of the motor
   (b) the current flowing in the armature

The back e.m.f. of a motor is also proportional to the magnetic flux of the poles and the speed of rotation. This can be expressed as

$$E \propto \Phi N$$

where

$$\Phi = \text{flux/pole in weber (W)}$$
$$N = \text{speed in rev/min}$$

This can be transposed for speed so that

$$N \propto \frac{E}{\Phi}$$

Later on, when speed control is considered, the fact that the speed is inversely proportional to the flux will prove very useful.

Generators produce an output voltage due to the movement of the armature conductors through the magnetic flux. As the armature conductors all carry current and have a resistance, it follows that they also produce a voltage drop. The actual terminal voltage of a generator is always slightly less than the generated voltage due to this voltage drop in the armature. This can be expressed as

$$U = E - I_a R_a$$

where

$$U = \text{terminal voltage (V)}$$
$$E = \text{induced (generated) e.m.f. (V)}$$
$$I_a = \text{armature current (A)}$$
$$R_a = \text{armature resistance (ohms)}$$

Example

What is the terminal voltage of a d.c. generator which has an armature resistance of 0.4 ohms and a current flowing through it of 4.75 A when it generates 250 V?

$$U = E - I_a R_a$$

filling in the details

$$U = 250 - (4.75 \times 0.4)$$
$$= 248.1 \text{ V}$$

Try this

1. A d.c. generator supplies 300 V when the armature current is 10.5 A and the resistance is 0.45 ohms. Calculate the generated e.m.f.

2. The generated voltage of a d.c. generator is 220 V and it has an armature resistance of 0.24 ohms. Calculate the terminal supply voltage when the armature current is
   (a) 10 A

   (b) 22 A

   (c) 40 A

The armature windings

In practice there are many coils on an armature and they are connected in one of two different ways. The number of coils and the way in which they are connected determines the number of segments on the commutator. The two types of winding are "Wave" and "Lap".

Wave windings

These are connected so that there are always two conducting paths in parallel and they tend to be used on smaller generators where high voltages are required at currents lower than an equivalent lap wound armature. Figure 1.21 shows a single coil connected to the commutator and Figure 1.22 shows the completed diagram of a wave wound armature.

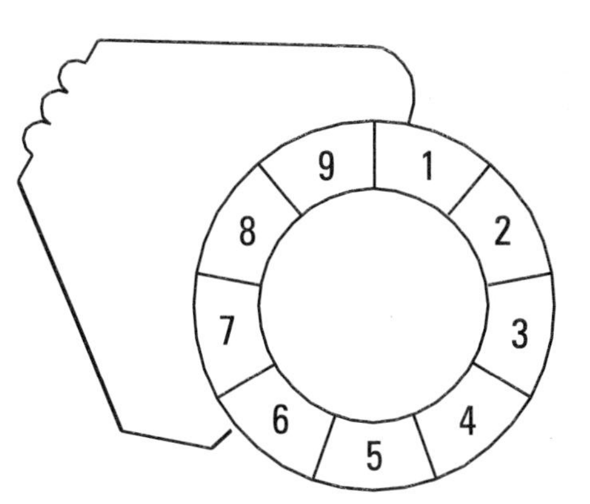

Figure 1.21 *Single winding from wave wound armature*

Figure 1.22 *A wave wound armature*

Lap windings

These tend to be low-voltage high current windings and have as many conducting paths in parallel as there are poles. For example a two-pole machine will have two parallel paths and a six pole machine will have six parallel paths. Figure 1.23 shows how the coils are connected to the commutator.

Figure 1.23 *A lap wound armature*

The field windings

The field windings in d.c. machines are basically coils of copper wire around a shaped iron pole. Figure 1.24 shows a representation of one.

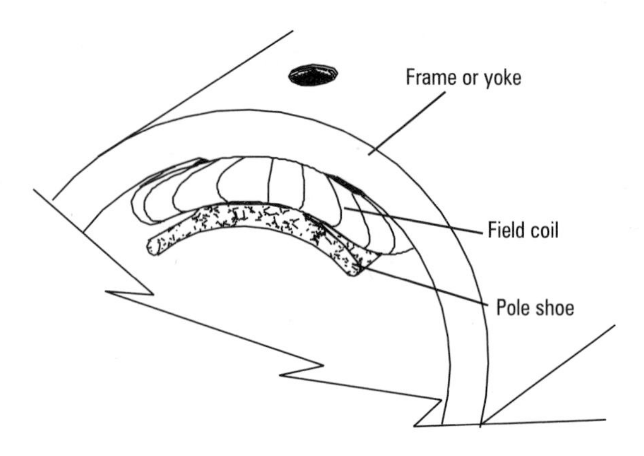

Figure 1.24 *The pole of a d.c. machine*

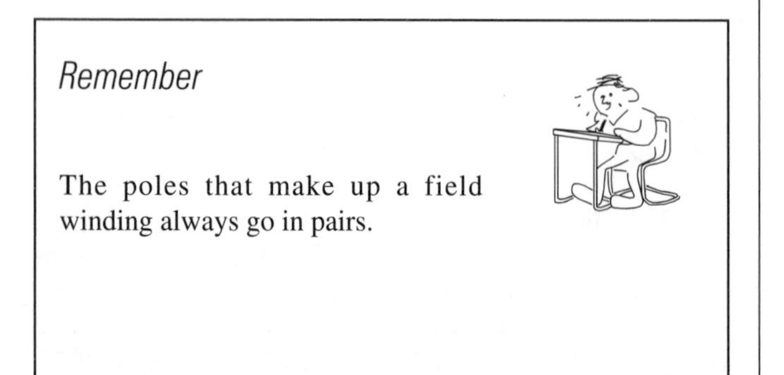

Remember

The poles that make up a field winding always go in pairs.

Induced voltages

There are three main factors that determine the amount of e.m.f. induced in a d.c. machine. These are

- the strength of magnetic flux
- the velocity of the conductors through the magnetic flux
- the length of conductor inside the magnetic field

The strength of magnetic flux

The magnetic strength of a pole is determined by the total flux it produces and its physical size. This is measured as flux density and can be expressed as

$$B = \frac{\Phi}{A}$$

where

Φ = total flux per pole (Wb)
A = the cross sectional area of the pole face (m^2)
B = magnetic flux density (T)

The velocity of conductors

It has already been seen that both motors and generators have voltages induced into their armatures due to the movement of the conductors. The instantaneous value of induced e.m.f. can be determined from

$$e = Blv$$

where

e = instantaneous value of induced e.m.f.
l = length of conductor in field (m)
v = velocity of conductor through field (m/s)

This shows that the velocity at which the conductor moves through the magnetic field is important. The velocity of the armature conductor is not determined by shaft speed but is of course related to it. This can best be explained by comparing two armatures which are rotating at the same shaft speed. If one has twice the diameter of the other the velocity of the conductors is much greater as it passes through the magnetic flux. The e.m.f. can be related to

$$E \propto \Phi n$$

where

Φ = total flux per pole (Wb)
E = induced e.m.f. (volts)
n = speed of rotation (rev/sec)

For each revolution that an armature rotates, each conductor cuts the flux at two points. One where the flux leaves the north pole, the other as the flux enters the south pole. This means that for each revolution the flux is taken at twice the number of pairs of poles (i.e. 2p).

In each revolution each conductor cuts across a magnetic flux equivalent to

$$2\,p\Phi\ \text{Wb}$$

where p = number of pairs of poles.

The total flux cut by one conductor in each second is

$$2\,p\Phi n$$

or

$$\frac{2\,p\Phi N}{60}$$

where N = speed of rotation in rev/min

So if it is given that one volt is induced when a flux of 1 Wb is cut per second,

the average e.m.f. per conductor = $\frac{2\,p\Phi N}{60}$ volts

or $2\,p\Phi n$ volts

The total number of conductors in series for each parallel path on an armature can be calculated from

$$\frac{Z}{a}$$

where
Z = number of armature conductors
a = number of parallel paths through the armature between brush connections

The number of parallel paths can be taken as 2 for a wave wound armature and 2p for a lap wound.

The total e.m.f. induced

$$E = \frac{2\,p\Phi N}{60} \times \frac{Z}{a}\ \text{volts}$$

Example

Calculate the induced e.m.f. in a 4 pole generator with a wave connected armature of 750 conductors when running at 1000 rev/min. The magnetic flux is 0.02 Wb.

$$E = \frac{2\,p\Phi N}{60} \times \frac{Z}{a}\ \text{volts}$$

Remember that on a wave wound armature $a = 2$:

$$E = \frac{2 \times 2 \times 0.02 \times 1000 \times 750}{60 \times 2}$$

$$= 500\ \text{V}$$

Try this

A d.c. generator is a 4-pole lap wound type. If it has a flux at each pole of 0.05 Wb and an armature with 600 conductors, what is the induced e.m.f. when it is working at 20 rev/s?

Torque

As shown on page 4, when a rotating movement is created a torque develops. The two factors that must always be considered when dealing with torque are:

- the force resulting from the rotary movement in Newtons (N)
- the radius at which the force acts, in metres (m)

Torque is measured in the result of these newton-metres (Nm)

Earlier in this chapter force on a conductor in a magnetic field was shown to be equal to:

$$F = BIl$$

For the purposes of d.c. machines this can be taken as

B = flux density of the poles (Φ)
I = current carried by the armature conductors (I_a)
l = length of armature conductor in the magnetic field

From this it can be seen that the force exerted on the conductors is proportional to the torque

$$F \propto T$$

As the length of conductor is governed by the armature construction, this remains a constant for each given machine. This means that the variable quantities are the magnetic flux of the poles of the machine, and the armature current.

Therefore $T \propto BI$

If either the armature current or flux in the poles is changed, the torque varies in proportion.

The rotating movement can be expressed in terms of doing work.

In this case

$$\text{work} = \text{force} \times \text{circumference}$$

$$= F \times 2\pi r$$

As

$$F \propto T$$

$$\text{work} = 2\pi T$$

As work can be expressed over a period of time

$$\text{Work/m} = 2\pi N T$$

where N = speed rev/min

As 1 Nm = 1 joule

and 1 joule = 1 watt/sec

$$\text{Power exerted} = \frac{2\pi N T}{60} \text{watts}$$

$$\text{Torque} = \frac{60 \times \text{power}}{2\pi N}$$

where N = rev/min

or

$$T = \frac{\text{output power}}{2\pi n}$$

where n = rev/sec

Example

A d.c. motor with an input power of 1.5 kW rotates at 800 rev/min and has an efficiency of 80%. Calculate

(a) the output power of the motor
(b) the torque developed by the motor

Answer

(a) Output of motor

$$= 1500 \times \frac{80}{100}$$

$$= 1200 \text{ watts}$$

(b) $$\text{Torque} = \frac{\text{output power} \times 60}{2\pi N}$$

$$= \frac{1200 \times 60}{2\pi \times 800}$$

$$= 14.32 \text{ Nm}$$

Efficiency

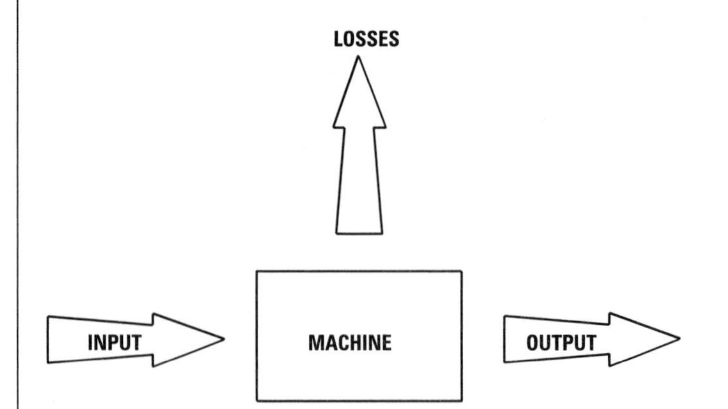

No machine is 100% efficient – there are always some losses. The losses can usually be divided into three areas

- the stator losses
- the armature losses
- and losses due to the mechanical movement of the machine.

The relationship between these can be seen below:

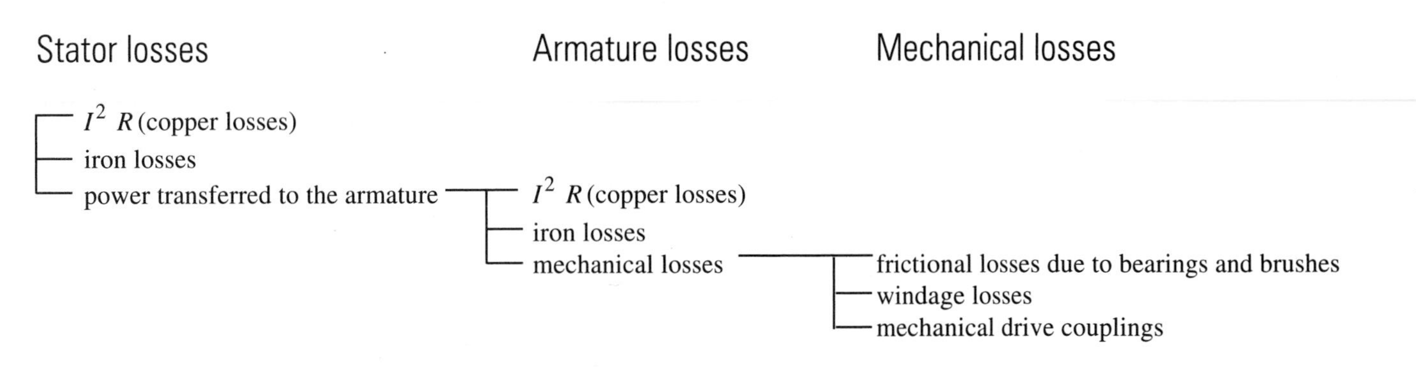

As can be seen from the chart above, the electrical losses are divided into two parts, the copper losses and the iron losses.

Copper losses

When a conductor carries a current, heat is produced due to its resistance. This heat is the power loss in the windings and is known as the I^2R losses. These apply to all conductors whether they are stationary or rotating. If an example is taken where a shunt motor has an armature resistance of 0.64 Ω and a current flowing of 3.25 A then this means that the copper losses for the armature are:

$$P_a = I^2 R$$
$$= 3.25^2 \times 0.64$$
$$= 6.76 \text{ watts}$$

The voltage drop across the brushes is usually about 2 V so if a current of 3.25 A is flowing in the armature a power loss of

$$P = UI$$
$$= 2 \times 3.25$$
$$= 6.5 \text{ W} \quad \text{occurs across the brushes.}$$

If the field winding has a resistance of 100 Ω when a current of 0.25 A is flowing the field losses are

$$P_f = I^2 R$$
$$= 0.25^2 \times 100$$
$$= 6.25 \text{ W}$$

The total copper losses for the motor are therefore

6.76 + 6.25 = 13.01 W

The total losses including the brushes is

13.01 + 6.5 = 19.51 W

Iron losses

The iron losses can be attributed to two areas related to the magnetic iron cores. These are the hysteresis losses and the eddy current losses.

Hysteresis

Where there is an alternating current flowing through a coil which creates a moving magnetic field in an iron core, a hysteresis loop is set up. By laminating these iron cores it is possible to reduce the loop to a minimum and keep the losses down.

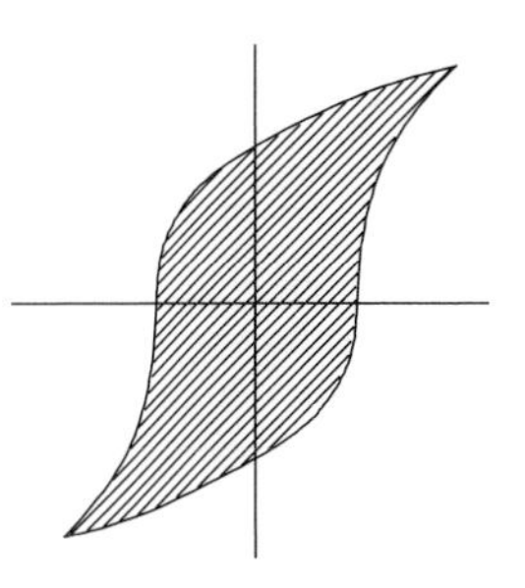

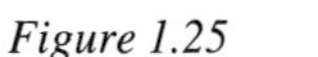

Figure 1.25 *High loss loop*

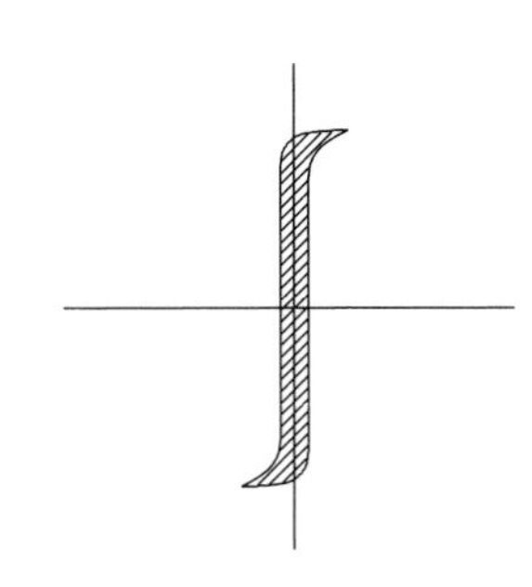

Figure 1.26 *Low loss loop*

The metal used for motor laminations is a magnetically soft material and is designed to have "good" low loss hysteresis properties.

Eddy currents

Eddy current losses are due to circulating currents set up in the iron laminations. As the iron used is an electrical conductor as well as being a magnetic conductor, currents are induced into it. As $P = I^2 R$ the greater the current induced into the iron core, the greater the power lost in it. This power is converted directly into heat which is not only a loss but can also damage the insulation of the windings. By using thin laminations with high electrical resistance the eddy current losses can be kept to a minimum.

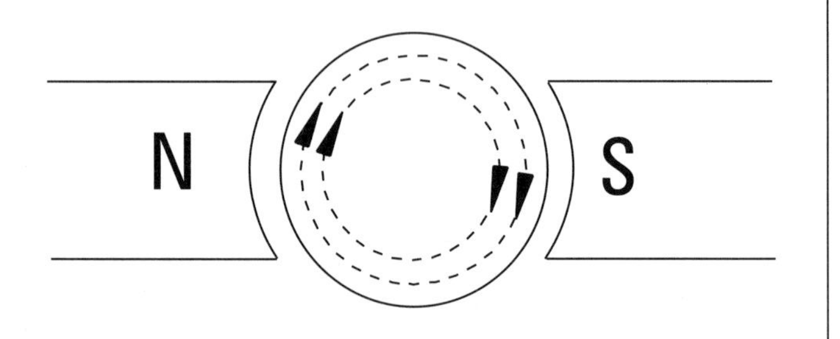

Figure 1.27 *Large metallic areas with low electrical resistance produce high eddy current losses.*

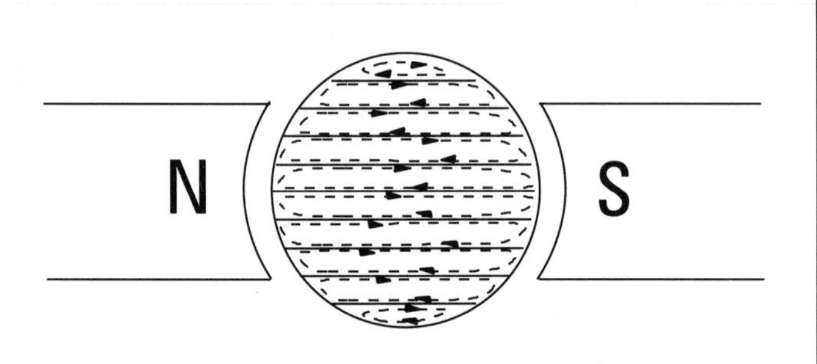

Figure 1.28 *Thin laminations with high electrical resistance produce low eddy current losses.*

Mechanical losses

As motors and generators are mechanical devices as well as electrical, there are always losses due to the movement created. These losses can be divided into bearing friction losses and windage losses. The losses due to bearing friction are proportional to the speed of the machine, the faster the machine rotates the greater the losses. The losses due to windage however, are proportional to the cube of the speed. Windage losses vary greatly from one machine to another depending on the cooling arrangement incorporated.

Total losses

The efficiency of a machine can be calculated from

$$\text{Efficiency } (\eta) = \frac{\text{OUTPUT power}}{\text{INPUT power}}$$

which can be shown as

$$\text{Efficiency } (\eta) = \frac{\text{input power} - \text{losses}}{\text{input power}}$$

REMEMBER

The answer to this equation must always be less than one. It is multiplied by 100 to give the efficiency as a percentage.

Example

A shunt motor connected to a 240 V d.c. supply has an armature resistance of 0.25 Ω and a field winding with a resistance of 245 Ω. Calculate

(a) the losses in
   i) the armature when a current of 25A is flowing in the armature and a voltage drop due to brush contact is a total of 1V
   ii) the field windings
(b) the total power loss due to iron and mechanical losses if these together account for 2.5% of the total motor input power
(c) the total power losses of the motor
(d) the maximum output power of the motor
(e) the efficiency of the motor as a percentage of the input power.

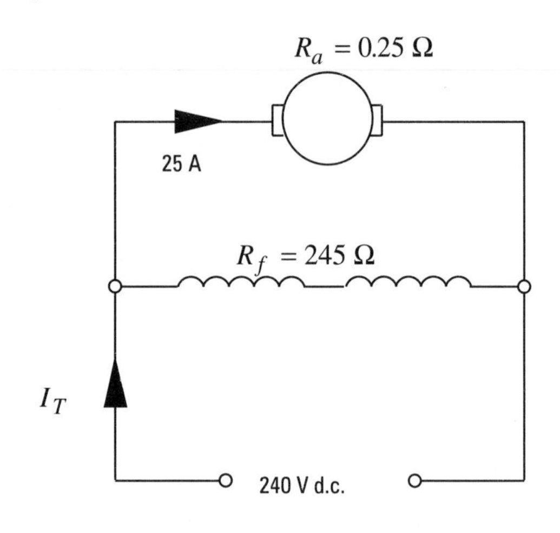

Figure 1.29

Answers:

(a) i) Armature copper losses

$$I_a^2 R_a = 25^2 \times 0.25 = 156.25 \text{ W}$$

Power loss due to brushes

$$U_b I_a = 1 \times 25 = 25 \text{ W}$$

Total power loss in armature circuit

$$= 156.25 + 25 = 181.25 \text{ W}$$

(a) ii) Field copper losses

$$P_f = I_f^{\,2} R_f$$

$$I_f = \frac{U_T}{R_f}$$

$$= \frac{240}{245}$$

$$= 0.979 \text{ A}$$

$$P_f = 0.979^2 \times 245$$

$$= 234.8 \text{ W}$$

(b) Total power input to motor

$$= U(I_a + I_f)$$

$$= 240 \times (25 + 0.979)$$

$$= 6234.96 \text{ W}$$

Iron and mechanical loss at 2.5%

$$6234.96 \times \frac{2.5}{100} = 155.874 \text{ W}$$

(c) Total power lost in the motor

$$181.25 + 234.8 + 155.874 = 571.924 \text{ W}$$

(d) Maximum output of motor

$$= 6234.96 - 571.924$$

$$= 5663.036 \text{ W}$$

(e) Efficiency of motor

$$= \frac{\text{input power} - \text{losses}}{\text{input power}} \times 100$$

$$= \frac{5663.036}{6234.96} \times 100$$

$$= 90.8\%$$

Try this

A shunt motor when connected to a 230 V d.c. supply develops a back e.m.f. of 224 V. The resistance of the armature and field windings are 0.4 Ω and 115 Ω respectively. The voltage drop across each brush is 1 volt.

Calculate
(a) the total power losses in the armature circuit
(b) the power losses in the field windings
(c) the maximum output of the motor if mechanical and windage losses are 3.5% of the input power to the motor
(d) the overall efficiency of the motor

Exercises

1. Calculate the force exerted on a 2.5 metre length of conductor when carrying a current of 12 A in a magnetic field at 90° to the pole faces which have a magnetic flux density of 0.75 T.

2. A 400 V d.c. motor has an armature resistance of 0.18 ohms.
   Calculate:
   (a) the value of induced e.m.f. when the armature current is 125 A
   (b) the value of the armature current when the induced e.m.f. is 382 V.

3. (a) Explain why the generated voltage is different to the terminal voltage in a d.c. generator.
   (b) A d.c. generator has an induced e.m.f. of 325 V when the armature current is 82.5 A and the terminal voltage is 282 V. Calculate:
   (i) the resistance of the armature
   (ii) the terminal voltage if the armature current increases to 90 A.

4. A 250 V d.c. shunt wound motor runs at 11 rev/s when the input current is 42 A. The armature circuit has a resistance of 0.25 Ω and a shunt field resistance of 62.5 Ω. Windage, friction and iron losses have a total of 1400 W. Calculate
   (a) the motor output
   (b) the efficiency of the motor
   (c) the torque developed on the shaft.

2

D.C. Motors

Complete the following to remind yourself of some important facts that you should remember from the previous chapter.

The amount of e.m.f. induced in a d.c. machine depends on the ____________ of the magnetic ________, the ________ of the conductors through the magnetic ________ and the __________ of conductor inside the magnetic __________.

Armature windings may be lap or ________ wound, and the latter has only __________ __________ paths.

Lap wound armatures are for __________ current, ______ voltage outputs.

Wave wound armatures are for high ________, low ______ outputs.

The cooling system of an electric motor will cause a ______ power loss, and bearings cause a ____________ power loss.

In the formula $E = \frac{2p\Phi N}{60} \times \frac{Z}{a}$

$Z =$ ________ of ____________ conductors, and $a =$ number of ________ __________.

Nearly all motors work on the basic principle that when a ________ __________ conductor is placed in a __________ __________ it experiences a __________. This ________ can be calculated using the formula

$$F = \quad \times \quad \times$$

On completion of this chapter you should be able to:

- describe the construction of a d.c. motor
- identify the characteristics of specific d.c. motors
- describe the methods of starting various d.c. motors
- draw diagrams to show methods of reversing the direction of rotation
- describe how the speed can be varied and controlled on various d.c. motors
- describe how the effects of armature reaction can be overcome
- describe how d.c. motors can be used on a.c. supplies

As with all electrical machines, d.c. motors can be divided into their mechanical and their electrical components. The mechanical ones consist of the case, bearings and drive arrangements, whereas the electrical side can be covered by the field and the armature. These could also be described as the stationary parts and the rotating parts.

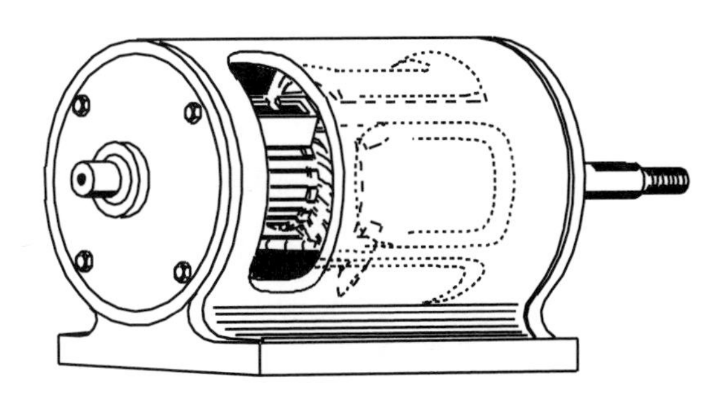

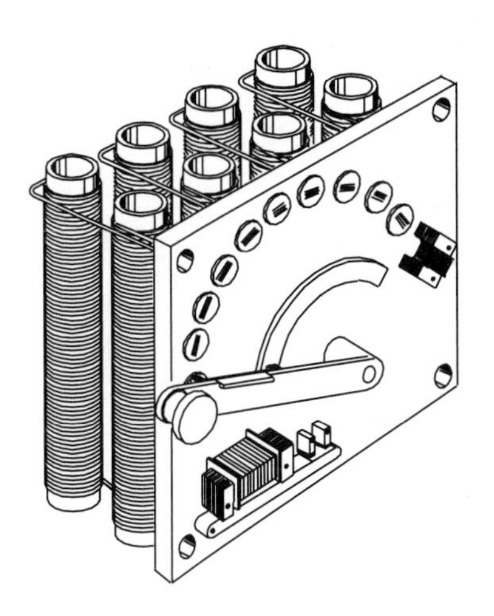

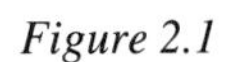

Figure 2.1 *d.c. motor and starter*

Construction

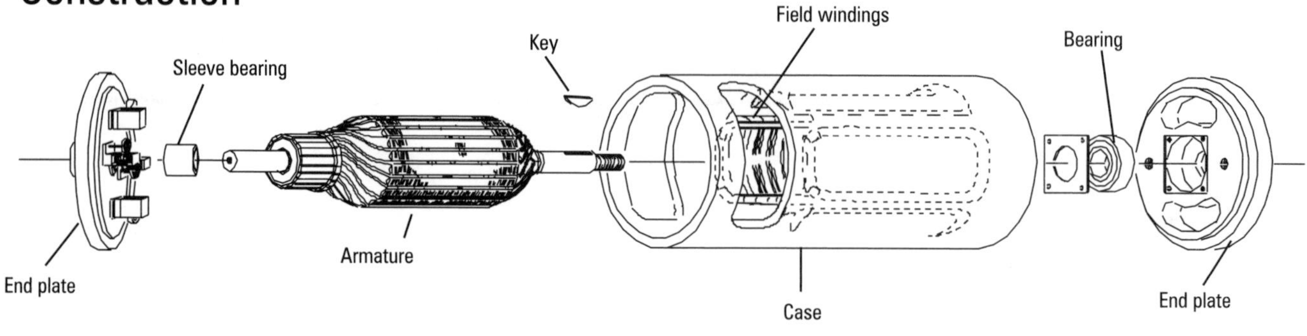

Figure 2.2 A d.c. motor used for starting a car engine

Figure 2.2 shows the main components of a typical d.c. motor which is used to start a car.

The field

The magnetic field in most motors is produced by electromagnets, unlike the motors in toy trains and cars that use permanent magnets. The magnetic field produced in the field windings has to be strong enough to go into the armature and interact with those windings. To ensure the strength of field is kept as strong as possible it must be carried in good magnetic conductors. The poles and case are constructed of iron which conducts the magnetic flux to the required parts of the motor. There will always be an air gap between the poles and the armature to allow it to rotate. As air is not a good conductor of magnetism the gap must be kept to a minimum. When current flows in the field windings a complete magnetic system of circuits is produced. Figure 2.3 shows typical magnetic circuits for a four-pole motor.

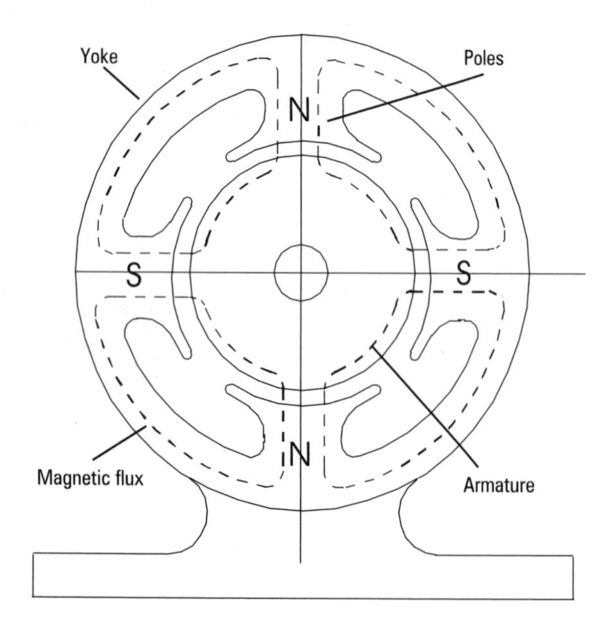

Figure 2.3 Magnetic circuit for a four pole motor

The magnetic flux flows in complete circuits going through one pole, round the frame, which when used in this way becomes the yoke, through the other pole and across to the first pole through the steel core of the armature to complete the magnetic circuit. The only air gaps in this circuit are between the armature and the pole shoes.

The poles on small motors are often made in one piece, with the core and shoe being shaped to take the winding. On larger motors the shoe is bolted to the pole core after the field coils are fitted. In these motors the pole core is usually made of solid iron and the pole shoe is laminated to reduce the effects of eddy currents being produced by the rotating armature conductors. The field windings are coils preformed to a shape so that they fit onto the pole core.

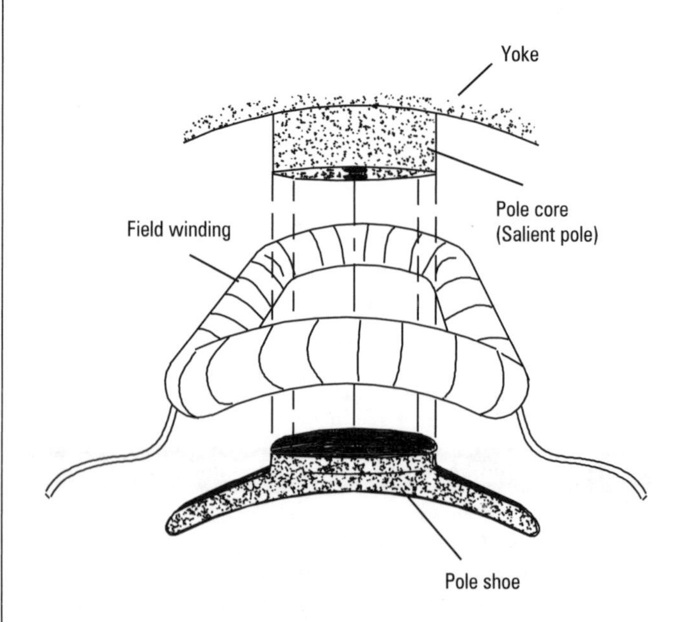

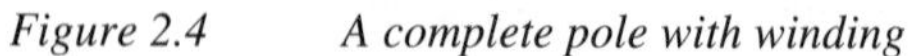

Figure 2.4 A complete pole with winding

The armature

The armature is made up of high quality magnetic alloy laminations mounted on a steel shaft with a commutator assembly at one end. The commutator consists of the appropriate number of copper segments all insulated from each other by mica based compositions. Each segment is dovetailed into an insulating ring to prevent movement due to heat and centrifugal forces. The whole commutator assembly is keyed onto the shaft.

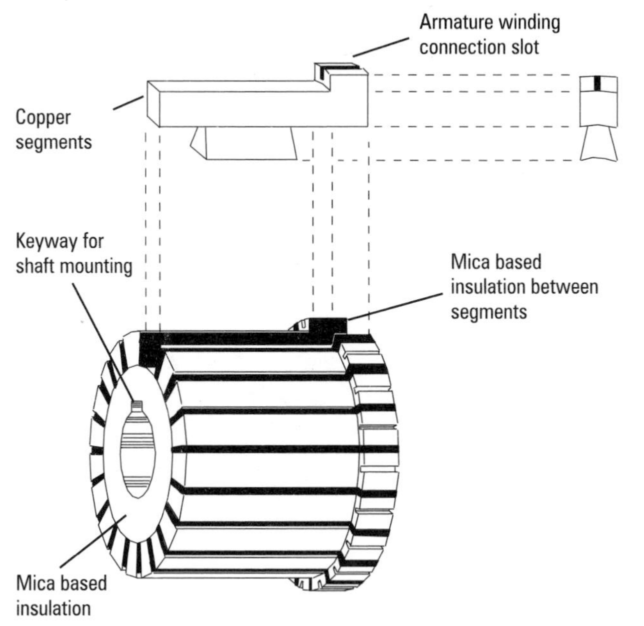

Figure 2.5 *A commutator assembly with one segment removed to show the make up*

The armature coils are usually preformed to the required shape before assembly. The coils are then fitted into the slots in the laminations and the ends terminated into the commutator connection lugs. The method used to insulate the windings from the laminations depends on the size and type of motor. This may be just a type of varnish on small motors or impregnated cloth on larger ones. After the armature is completely assembled and all of the windings are securely fitted, it must be checked to ensure it is electrically and mechanically sound. Insulation tests at the appropriate voltages must be carried out and the results recorded. The whole assembly must also be checked to ensure it is in alignment and balanced.

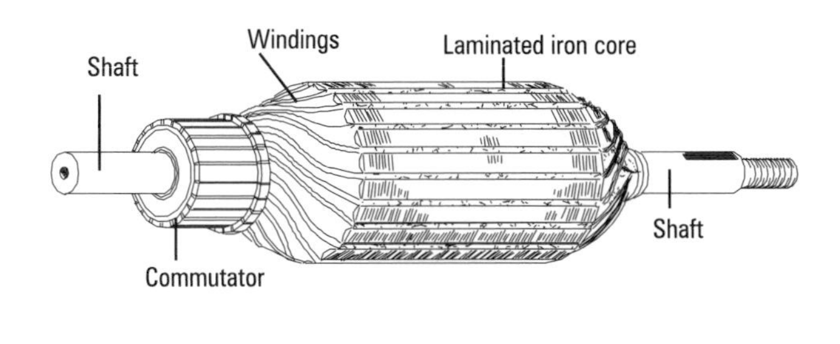

Figure 2.6 *Complete armature assembly*

Brush connections

Electrical connection is made to the commutator through carbon or graphite brushes. Each brush is held in place by a box arrangement which is fitted securely to the motor frame. The pressure that ensures good electrical contact is produced by a flat spring pushing onto the top of each brush. As the brush wears down through use, the spring is designed to keep a constant pressure. The brush is connected to the electrical circuit by a flexible "pigtail" fitted into the top of the brush and terminated onto the windings away from any movement.

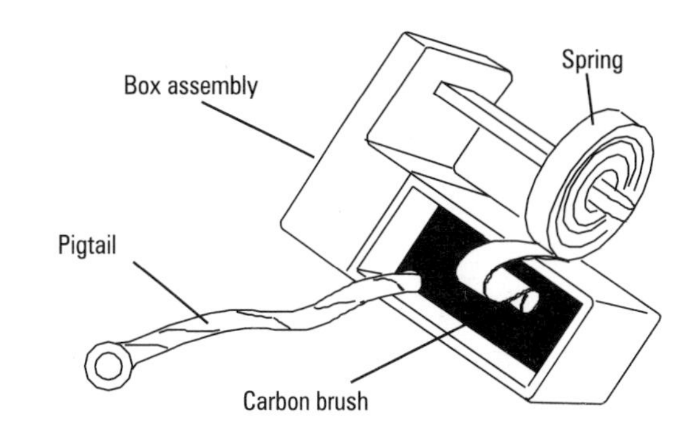

Figure 2.7 *A brush in a box*

Try this

Before continuing, list below applications of d.c. motors.

Types of motor available

In the first chapter the permanent magnet motor was referred to for driving small electric toys. Although this is used extensively for that purpose it has very few applications in industry. In general d.c. motors take their name from the connection arrangement of the field and armature windings. Examples of these are:

- the series motor
- the shunt motor
- the compound motor

These will now be dealt with in detail in this chapter.

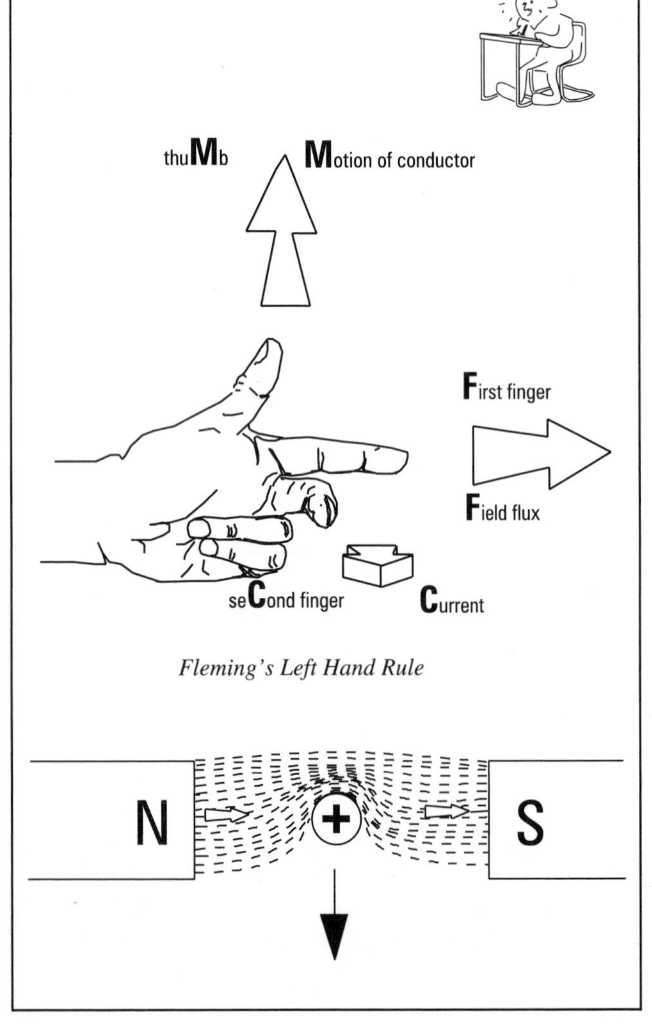

Fleming's Left Hand Rule

The series motor

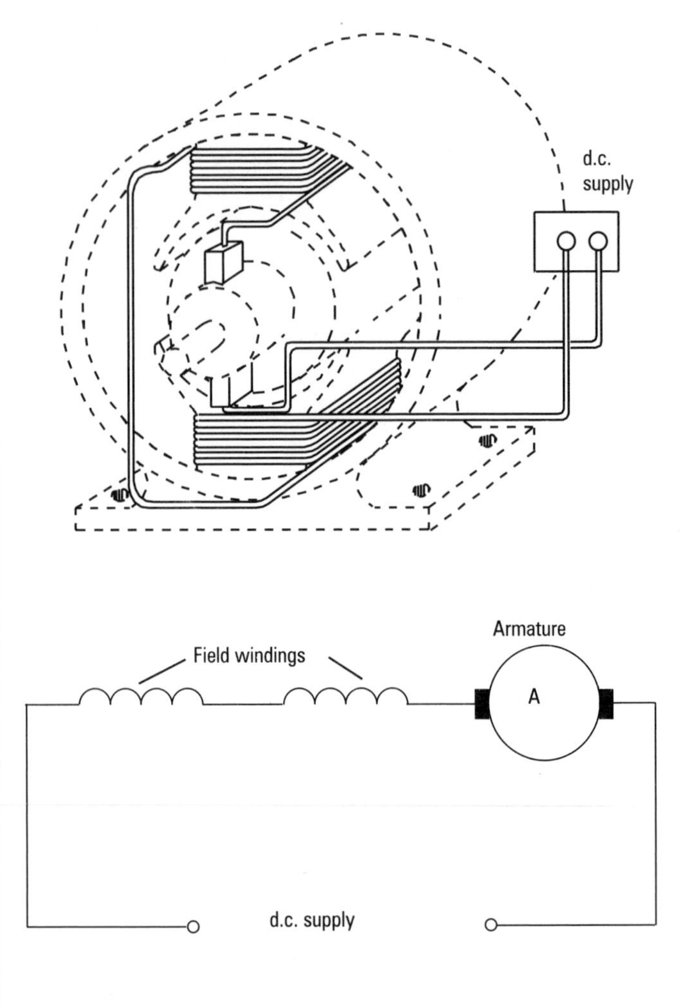

Figure 2.8 *A connection diagram, and circuit diagram, for a series wound motor*

The first motor to consider in some detail is the series type. As the name implies, the field and armature windings are all connected in series as shown in Figure 2.8.

In a series motor the current flowing in the armature also flows through the field windings, so when the motor is put under load, the armature and field currents increase. This means that the magnetic flux becomes stronger and more torque is produced. When the load is reduced, the magnetic flux becomes weaker and the speed increases. In theory, if a series motor was left with no load connected, it would continue to increase speed until it destroyed itself.

This type of motor is used where large starting torques are required, such as on traction engines and cranes. Figures 2.9a and b show the relationships between speed/torque, and torque/load, for this type of motor.

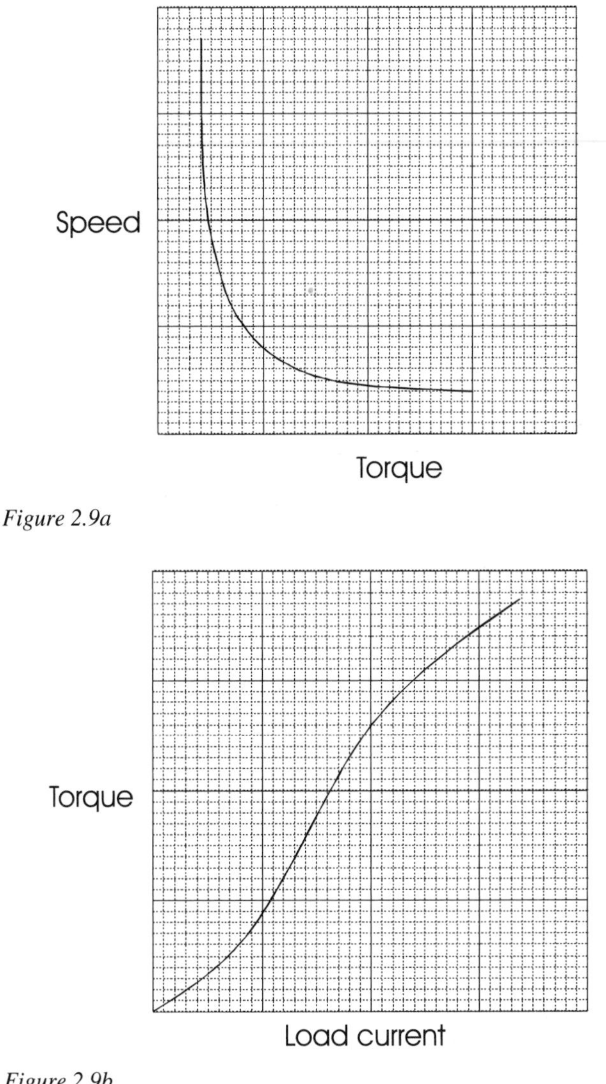

Figure 2.9a

Figure 2.9b

Speed control of a series motor

The speed on this type of motor can be controlled by using "diverter" resistors across either the field or the armature windings. The field diverter shunts off some of the current from the field windings making the field weaker and therefore increasing the speed. When a diverter resistor is used across the armature the current in the field windings is increased and the motor is slowed down.

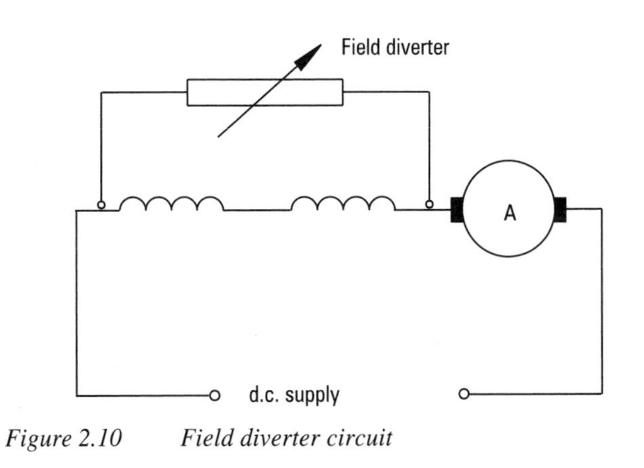

Figure 2.10 Field diverter circuit

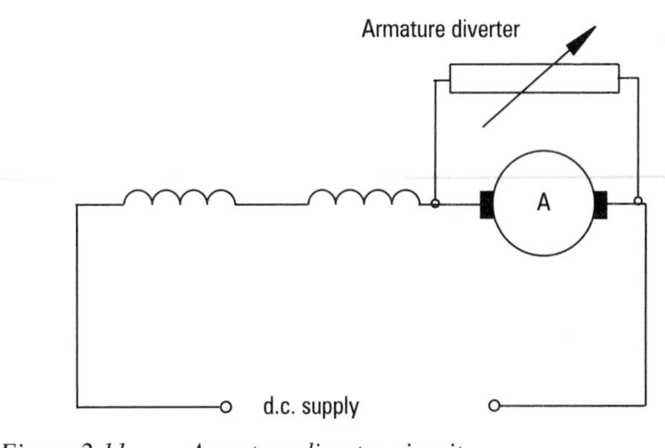

Figure 2.11 Armature diverter circuit

The problem when using any variable resistor in this way is the current the resistor has to carry. Each resistor must be capable of carrying the maximum current for the speed range it is designed for. This means that they are often expensive and bulky and produce heat which lowers the overall efficiency of the motor. Some motors have tapped field windings so that the field strength can be adjusted without the use of resistors (Figure 2.12).

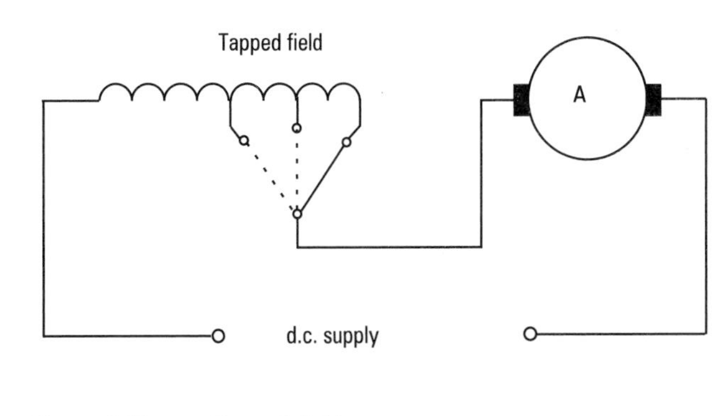

Figure 2.12 Tapped field

Starting

Starting a d.c. series motor must be carried out using the correct equipment. When the motor is stationary no back e.m.f. is being produced so it will take the maximum current from the supply. To reduce this current while the motor is building up its speed and its back e.m.f., a number of resistors are connected in series with it. These reduce the supply voltage and hence the current drawn by the motor. As the speed increases the resistors are shorted out so that when the motor reaches full speed it is connected straight across the supply. At this stage the back e.m.f. is at maximum so the resistors are no longer required.

The starter used for this is also designed to give overcurrent protection and "no volt" release. Overcurrent protection is monitored by a coil connected in series with the motor, so that if the supply current increases beyond a predetermined value, the coil becomes magnetically energised. At this stage an armature is attracted to the coil and a pair of contacts are shorted out. These contacts are connected across the electromagnet that holds the starting arm in the run position. When the electromagnet becomes de-energised the starting arm automatically returns to the "off" position. This same electromagnet also acts as the "no volt" protection, so should there be a power failure the machine will automatically switch off until it is manually restarted. This type of starter is often referred to as a "faceplate starter" due to its flat vertically mounted construction. Figure 2.13 shows a circuit diagram for a faceplate starter for a series motor.

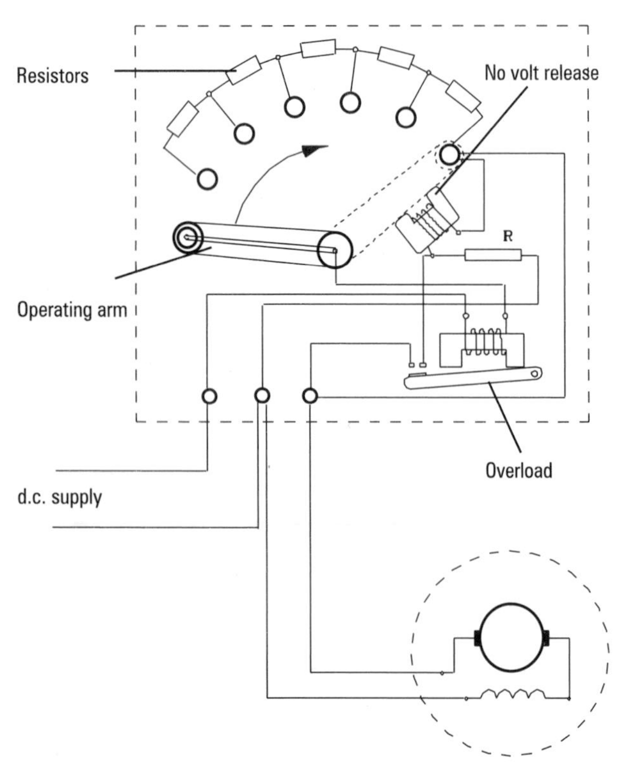

Figure 2.13 *Faceplate starter for a series motor*

Reversing the direction of rotation

In Chapter 1, it was seen that when a single current-carrying conductor was placed in a magnetic field and the direction of the current was reversed then the direction of the movement in the conductor was also reversed. However, when BOTH the magnetic field and current were reversed then the movement on the conductor remained the same as before the changes were made. D.C. motors respond in a similar way. If the supply polarity is changed, the motor will still rotate in the same direction as before. This is because in effect both the armature and field windings have been reversed. If either the field or the armature connections are changed over the motor will rotate in the opposite direction, but if both are changed the direction of rotation stays the same. Figure 2.14 shows the circuit diagrams for changing either the field or armature connections. Refer to Figure 2.8 for the original connections and then compare it with the two circuits in Figure 2.14.

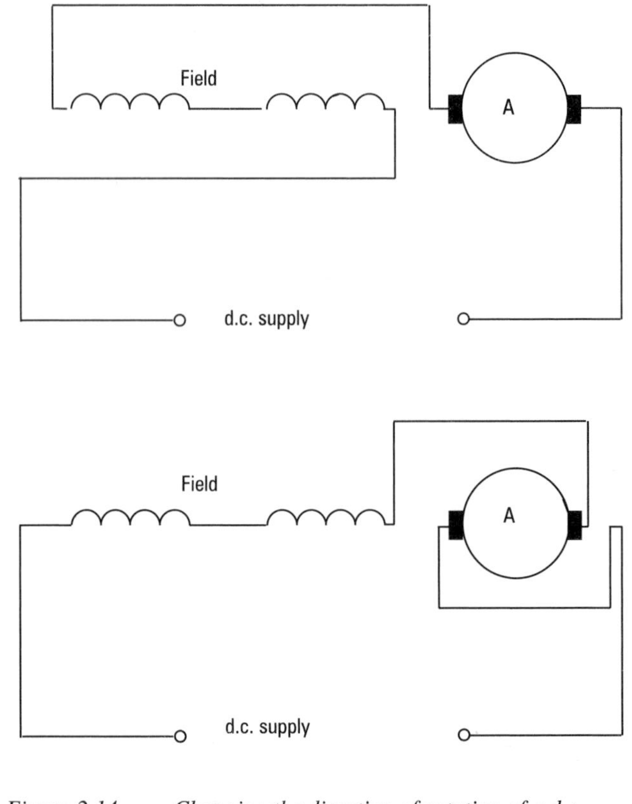

Figure 2.14 *Changing the direction of rotation of a d.c. series motor*

Series motor summary

Windings
Armature and field windings connected in series.

Load type
On high load speed decreases and torque increases.

Starting
Variable resistances connected in series.

Speed control
By variable resistors connected either across the armature or field windings.

Reversing rotation
Changing either the armature or field connections but **NOT** both.

The shunt motor

This is a motor where the field and armature windings are connected in parallel, as shown in Figure 2.15.

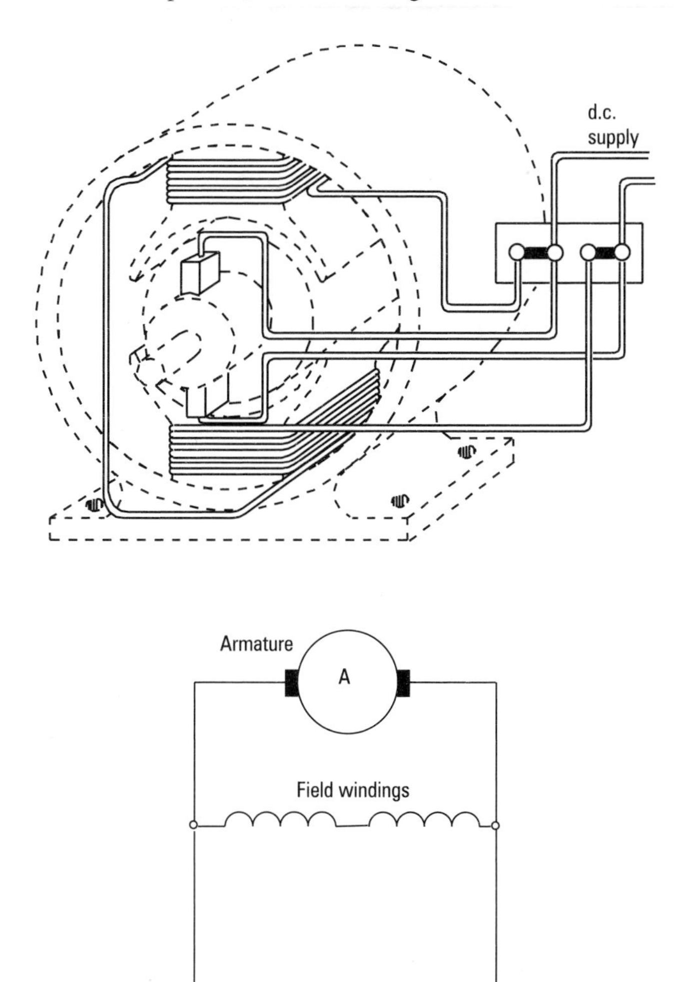

Figure 2.15 *Connection diagram and circuit diagram of a shunt wound motor*

When a motor is connected in this way the field windings receive the full supply voltage across them. As the supply is constant the field strength is constant and therefore the motor speed is fairly constant. As the field strength does not vary Φ is constant and the speed (n) is proportional to the back e.m.f. (E). As the load is increased the armature current and voltage drop also increase producing a slight reduction in motor speed. This is due to the reduction in magnetic flux as the armature reaction takes place. Figures 2.16 a and b show the characteristics for this type of motor.

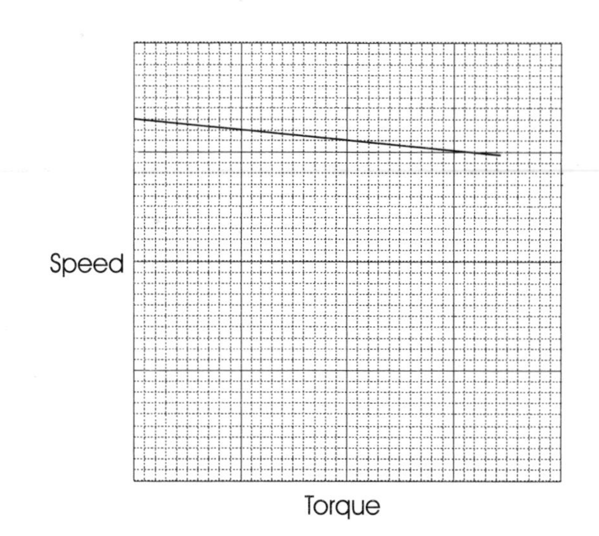

Figure 2.16a

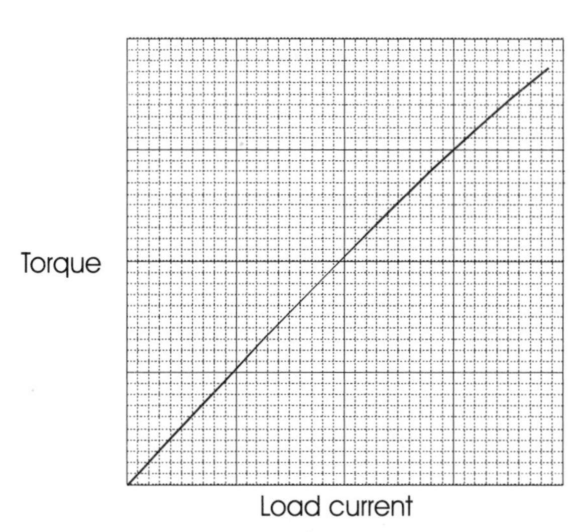

Figure 2.16b

Speed control of a shunt motor

Speed control of shunt motors is obtained by connecting a variable resistor in series with the field windings, as shown in Figure 2.17. When this is adjusted to increase the resistance, the current is reduced and so the magnetic field strength is weakened. To produce the same back e.m.f. the motor has to run faster so the speed increases. Shunt motors can have their speed varied using resistors in this way within a range of about 3 to 1 without weakening the poles too much.

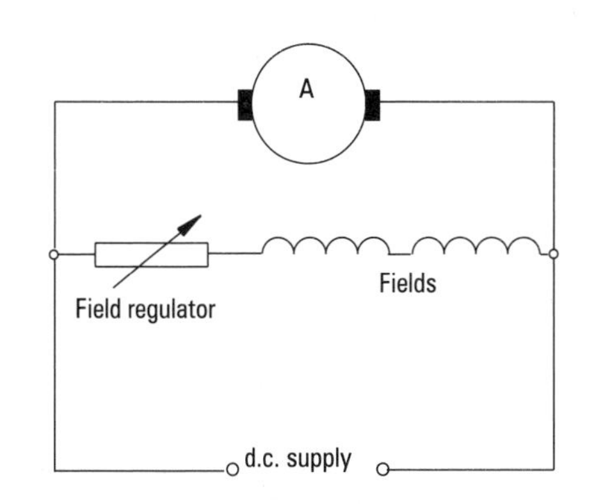

Figure 2.17 *Circuit diagram of shunt motor with field regulator*

Try this

When starting a d.c. motor explain why it is important to remember there is no back e.m.f. when the motor is stationary.

Starting

The starting procedure for a shunt connected motor is similar to that of the series motor. In both types a faceplate starter is used to add resistance into the circuit during the starting process. When starting the shunt connected motor the resistances are connected into the armature circuit only as this is where the high starting currents will be. When the armature is rotating at full speed and the maximum back e.m.f. is being produced, the resistors are shorted out. Overcurrent and no voltage protection are similar to those on the series type starter. Figure 2.18 shows a circuit diagram for a faceplate starter for a shunt connected motor.

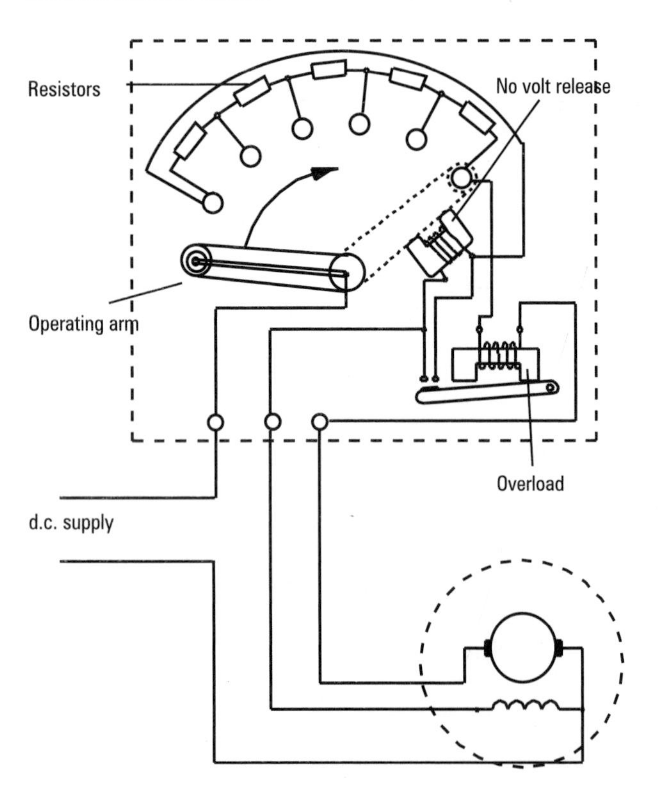

Figure 2.18 Circuit diagram for shunt connected motor

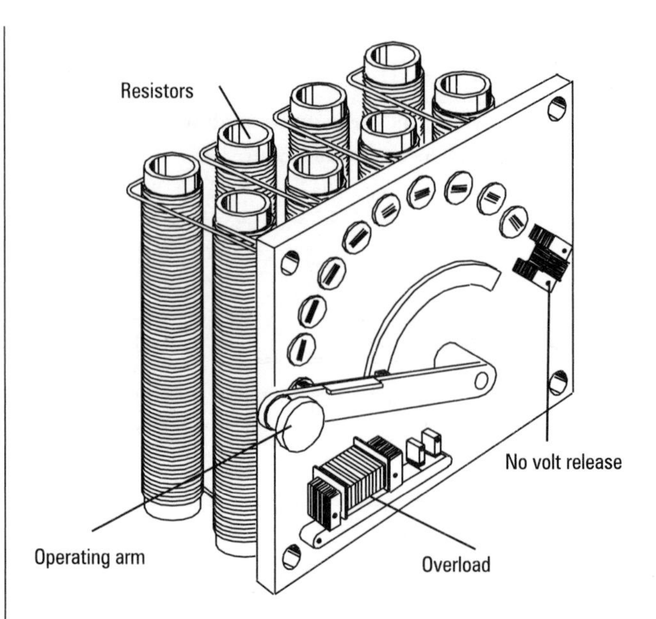

Figure 2.19 Construction of a faceplate starter

Reversing the direction of rotation

Like the series motor the shunt connected motor has to have either the connections to the field windings or armature windings changed over, not both. Figure 2.20 shows circuit diagrams of how this can be carried out. See Figure 2.17 for the original circuit.

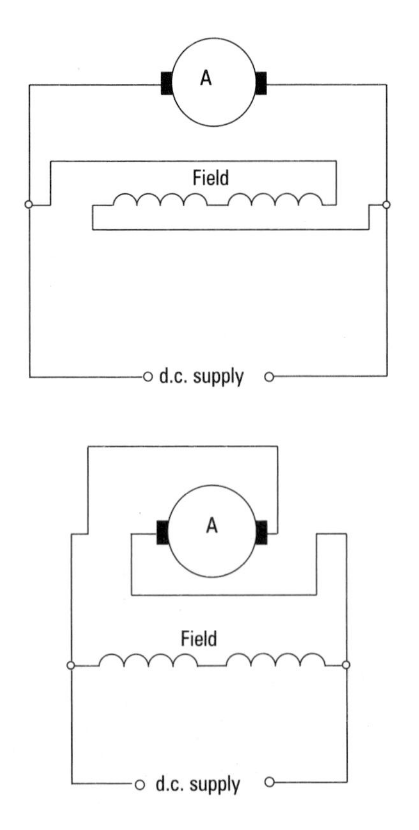

Figure 2.20 Changing the direction of rotation of a d.c. shunt motor

Shunt motor summary

Windings
Armature and field windings are connected in parallel across the supply.

Load type
As the load increases the torque increases and the speed slightly decreases.

Starting
Variable resistances are connected into the armature circuit.

Speed control
By using variable resistors connected in series with the field windings.

Reversing rotation
Changing either the armature or field connections but **NOT** both.

Example
Assume that a d.c. motor with an armature resistance of 0.2 Ω is switched directly on to a 220 V d.c. supply, without a starter.

$$U = E + I_a R_a$$

$$\therefore I_a = \frac{U - E}{R_a}$$

At the instant of switching on the motor will be stationary, so the back e.m.f. will be zero.

$$\therefore I_a = \frac{U - 0}{R_a} = \frac{220 - 0}{0.2} = 1100 \text{ A}$$

Try this
A 200 V d.c. motor has an armature resistance of 0.16 Ω. Calculate the starting current if no starting resistor is connected into the armature circuit. Ignore the field resistance.

The compound motor

This type of motor has both shunt and series connected field windings, as shown in Figures 2.21a, 2.21b and 2.21c.

The characteristics of the motor depend on which set of field windings is the strongest. If the series winding is used to increase the magnetic pole strength as the load current increases, this is known as cumulative compounding. This type of motor has characteristics between the series and shunt types, giving high starting torques with safe no-load speeds. This makes cumulative compounded motors ideal for heavy intermittent loads such as lifts and hoists.

There are two methods of connecting the shunt windings of a compound motor. Figure 2.21a shows the long shunt connections, whereas Figure 2.21b is connected in short shunt.

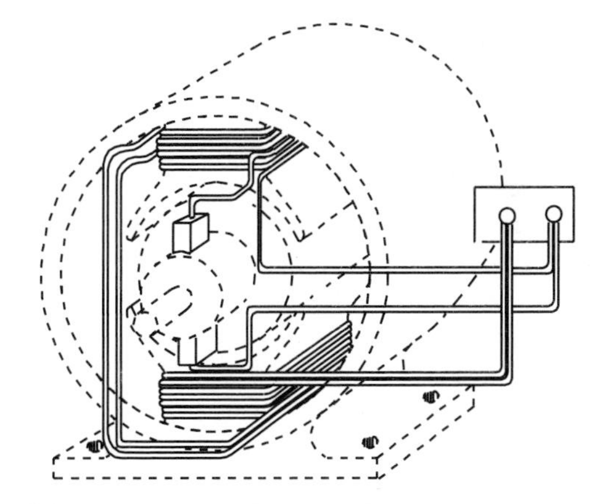

Figure 2.21a Compound connected motor

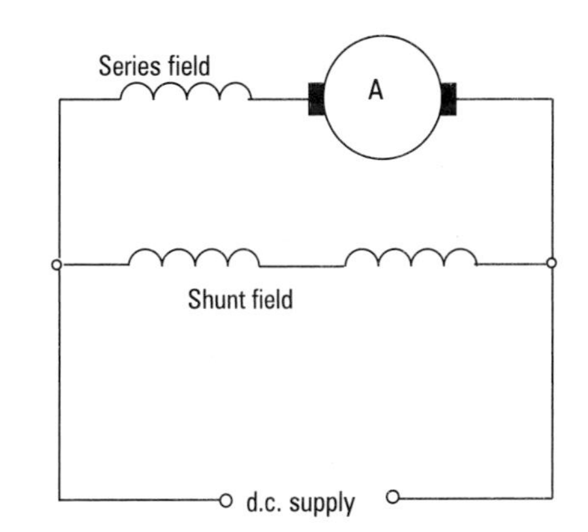

Figure 2.21b Long shunt connected

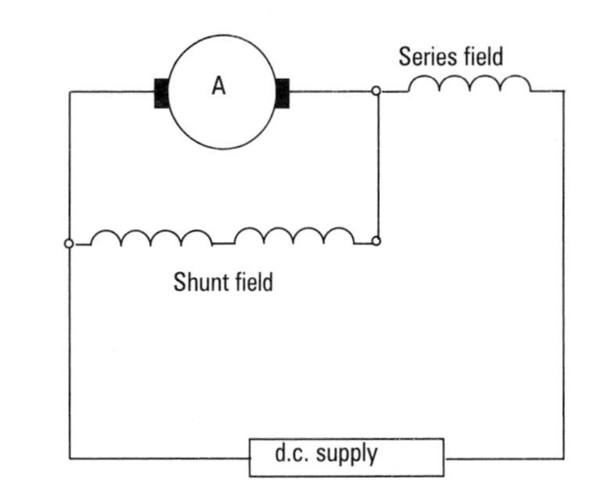

Figure 2.21c Long shunt connected

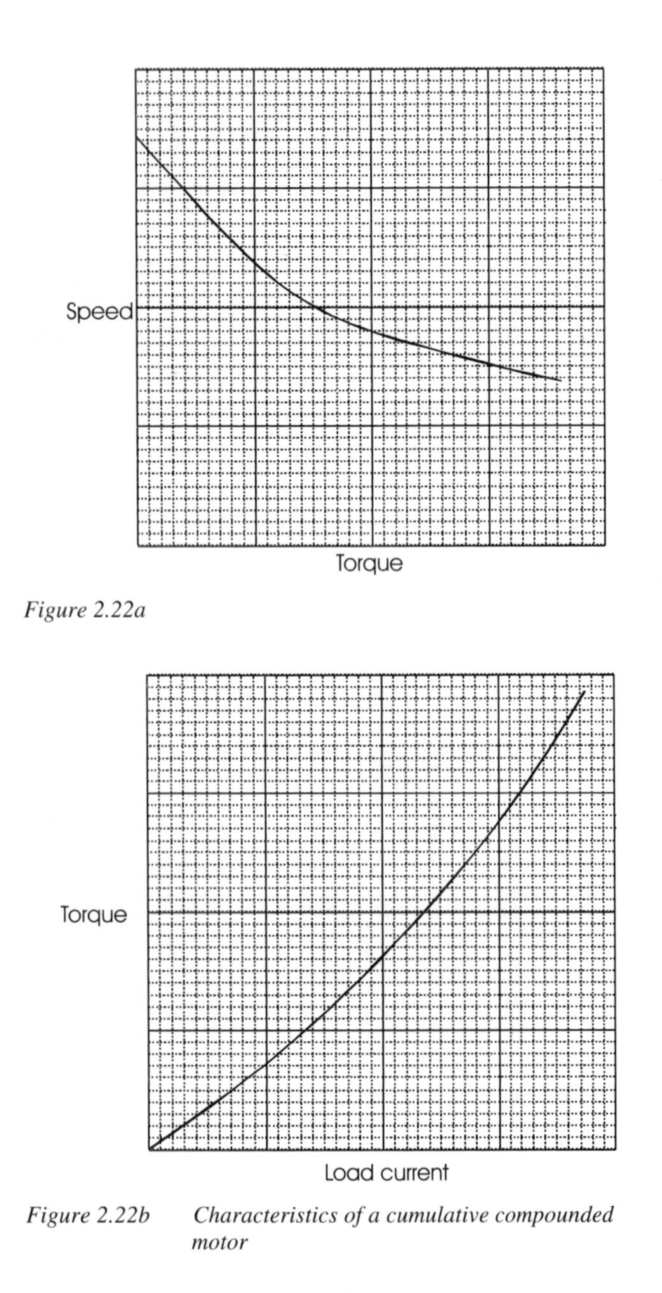

Figure 2.22a

Figure 2.22b *Characteristics of a cumulative compounded motor*

Speed control of a compound motor

When the series winding is used to weaken the magnetic pole strength as the load current increases, this is a differentially compounded motor. In these the magnetic flux produced by the series field winding opposes and has the effect of weakening the magnetic flux set up by the parallel field winding, which then increases the motor speed as load current increases. These motors have a tendency to become unstable as the differential compound effect can cause the speed of the motor to rise to uncontrollable levels. This instability means that this type of machine is seldom used in practice. It is important to know what can result so that motors are not accidentally connected in this way.

The speed of cumulative compound motors can be controlled by either series or shunt connected variable resistors. In some circumstances both series and shunt are used.

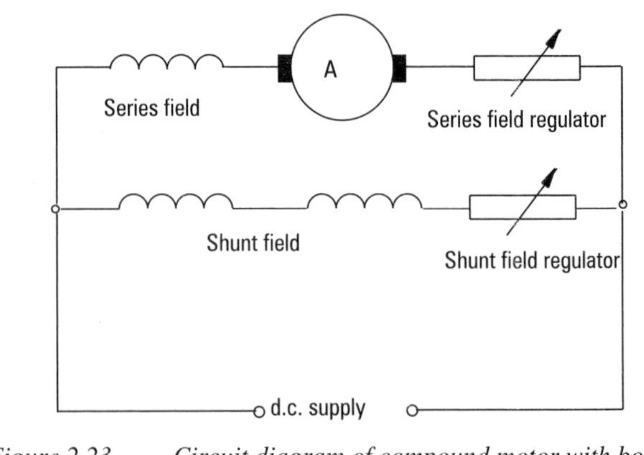

Figure 2.23 *Circuit diagram of compound motor with both series and shunt connected variable resistors*

Reversing the direction of rotation

To reverse the compound motor the connections to the armature are changed as shown in Figure 2.24.

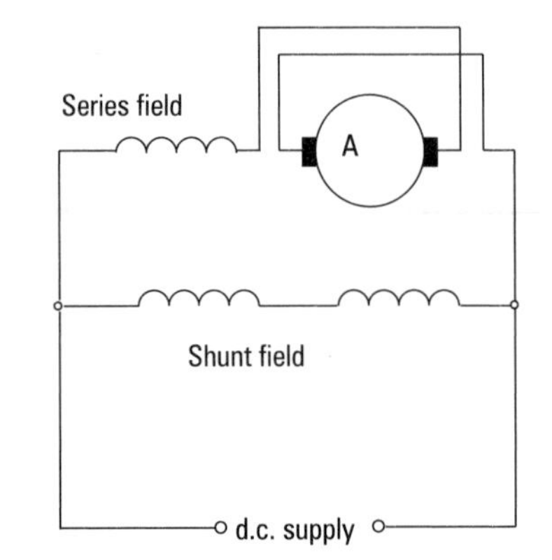

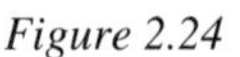

Figure 2.24

To reverse the direction using the field windings, both sets have to be reversed as shown in Figure 2.25.

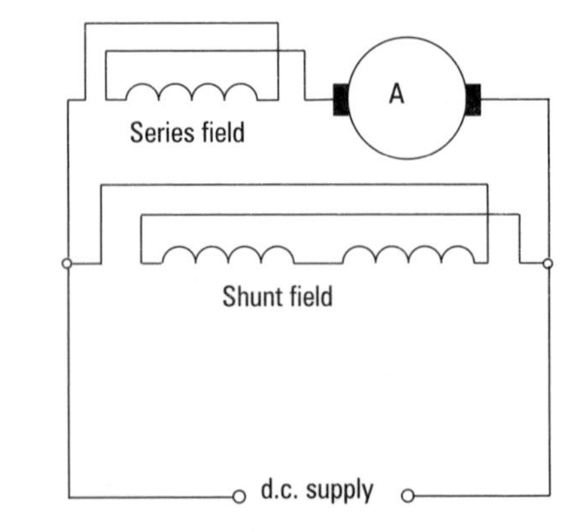

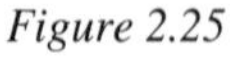

Figure 2.25

Note: If **only one field** is reversed the characteristics of the motors will be changed (i.e. wrong connection of the fields will alter the fields from cumulative to differential or vice versa).

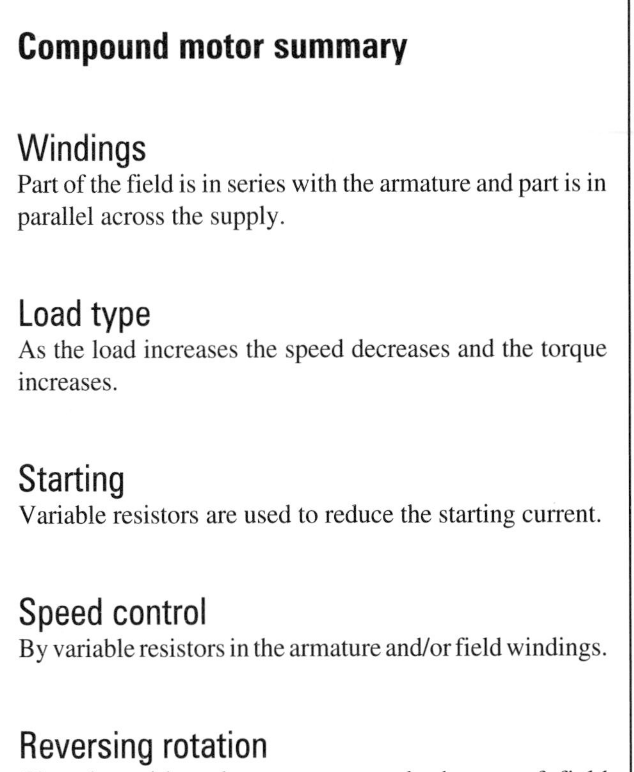

Compound motor summary

Windings
Part of the field is in series with the armature and part is in parallel across the supply.

Load type
As the load increases the speed decreases and the torque increases.

Starting
Variable resistors are used to reduce the starting current.

Speed control
By variable resistors in the armature and/or field windings.

Reversing rotation
Changing either the armature or both sets of field connections but **NOT** both.

Armature reaction

When the armature in a motor rotates the generated (back) e.m.f. opposes that supplied. When the motor is loaded this has the effect of distorting the angle of magnetic flux backwards as shown in Figure 2.26.

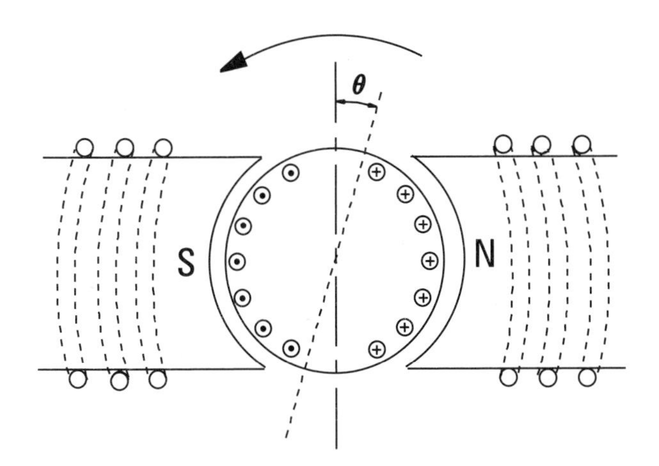

Figure 2.26 Diagram showing the change in magnetic axis

As the brushes are normally situated on the magnetic axis to ensure the current is reversed at the appropriate time, the brush position must be adjusted when the motor is run under load. When the brush position is changed this tends to de-magnetise some of the field conductors, and this reduction in flux causes an increase in speed. As the angle of the magnetic axis varies with load, in theory the position of the brushes should be adjusted every time the load changes. On small motors where the load is fairly constant the brush position can be set and checked when maintenance is carried out.

To overcome the problems associated with heavy loads on d.c. motors, a compensating winding, or interpole, is fitted to the yoke. This is a small pole situated between the main poles of the field windings. It is constructed in a similar way to the other poles but the coil is connected in series with the armature.

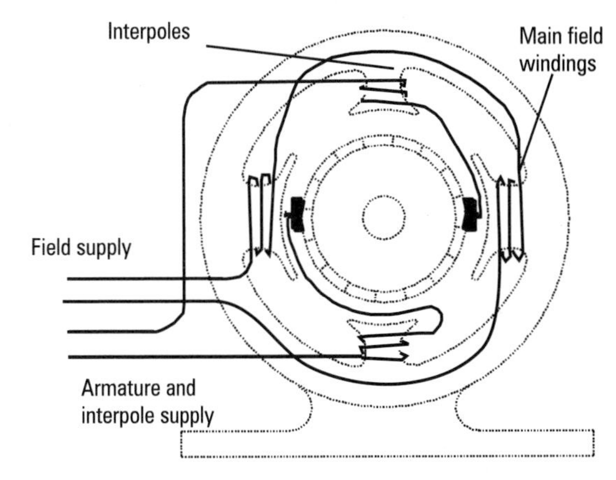

Figure 2.27 Connection diagram of interpoles

The combined effect of the currents in the armature and interpole bends the flux in the air gap, between the armature and field poles, overcoming the effect of armature reaction without loss of flux density. The use of interpoles tends to be restricted to larger machines and those used for special purposes, due to the extra cost involved.

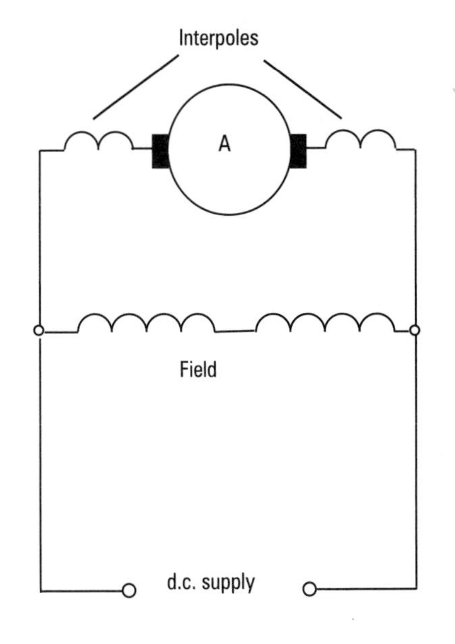

Figure 2.28 Circuit diagram showing the connection of interpoles

Commutation

As has already been shown the generated e.m.f. in an armature is alternating even though the motor is supplied with d.c. As the armature rotates the commutator is continually reversing the current and hence the magnetic field. The rotation of the commutator under the brushes means that segments are continually being shorted out (Figure 2.29).

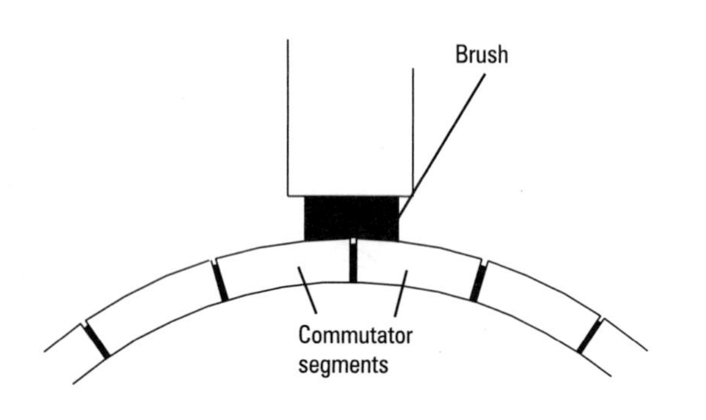

Figure 2.29 *Commutator segments shorted across by brush*

If, for simplicity, a lap wound armature is considered, each winding is connected across adjacent segments on the commutator. As the armature rotates each winding carries current and produces a magnetic field. At the instant the two segments are shorted out by the brush being across the insulated gap, the winding is also shorted out (Figure 2.30).

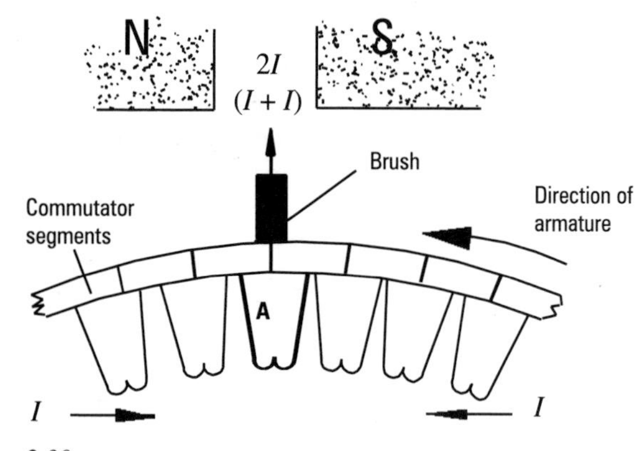

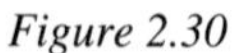

Figure 2.30

This means that the magnetic field in that section of the armature (**A**) suddenly collapses and a high e.m.f. is induced. As the commutator is still rotating the induced e.m.f. is discharged across the brush and the segment that it has just left. If the brushes were made of a good electrical conductor the spark would be very hot and burn away the edge of the segment and back edge of the brush.

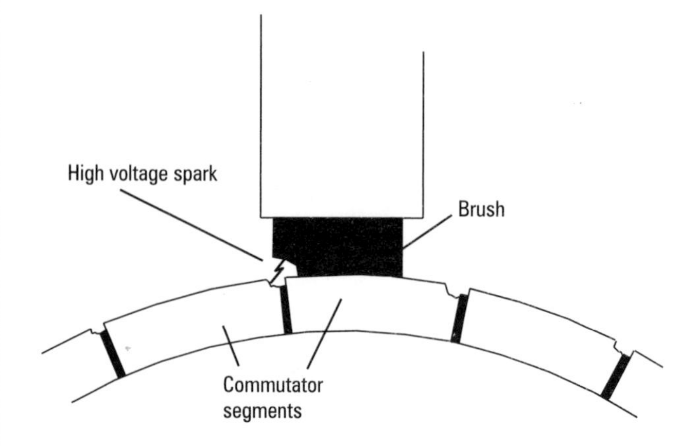

Figure 2.31 *Burning of brush and segments due to incorrect brushes*

As the armature continues to rotate the current flows in that winding in the opposite direction and a new magnetic field builds up. To overcome this effect of commutation, brushes are made of carbon compounds that have relatively high resistances. This allows the high induced e.m.f. to discharge comparatively slowly and little or no spark is produced.

The motor designer will calculate the correct composition of carbon for the brushes of each machine. Care must always be taken when replacing brushes to ensure that the effects of commutation are kept to a minimum.

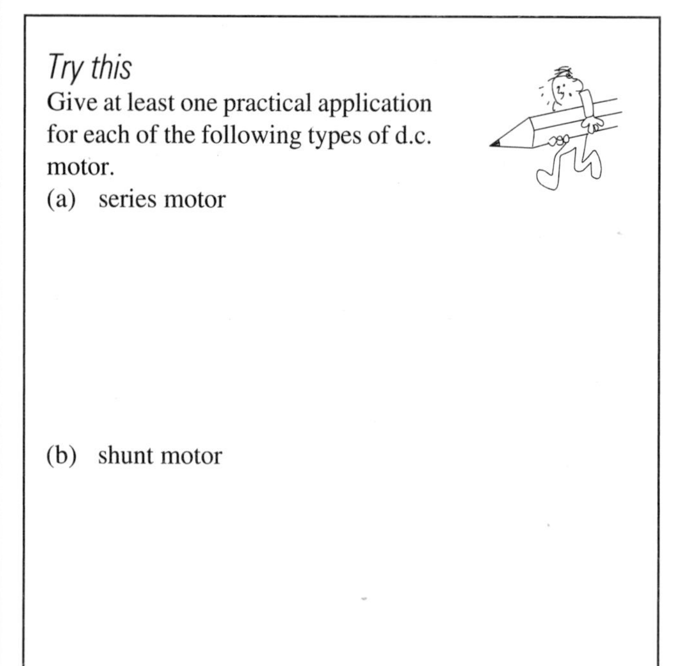

Try this

Give at least one practical application for each of the following types of d.c. motor.

(a) series motor

(b) shunt motor

(c) compound motor

Brush contact

The brush must not only be made of the appropriate composition of carbon but pressure applied to the brush to ensure good contact is also important. The brush has to make good electrical contact with the surface of the commutator which is moving at high speed. If the pressure is too great, the friction losses become a problem. If, however, the pressure is not great enough, the electrical losses become greater. Figure 2.32 shows the relationship between brush pressure and power loss.

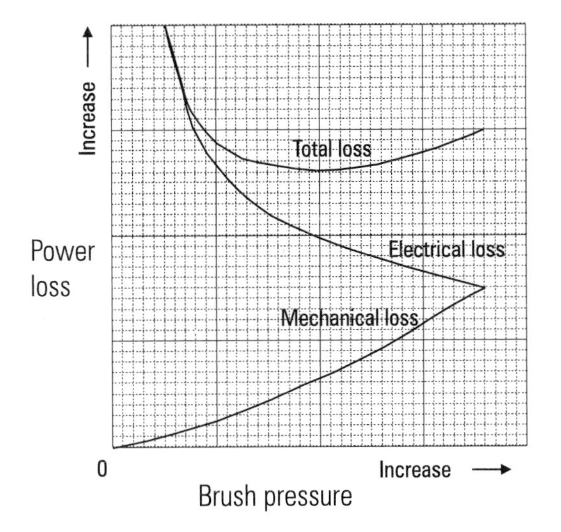

Figure 2.32

D.C. motors used on a.c. supplies

With the development of sophisticated electronics, d.c. motors of all types can be used on a.c. supplies. Packages have been developed that allow d.c. motors to be completely controlled so that constant speed or speed control can be an integral part. An example of a constant speed circuit is shown in Chapter 4 where it is used on a series "universal" motor. When speed control is to be used, a microprocessor can be incorporated and any desired speed/torque curve can be achieved for a given motor.

A package for a d.c. shunt motor will normally consist of two parts, as shown in the simplified Figure 2.33.

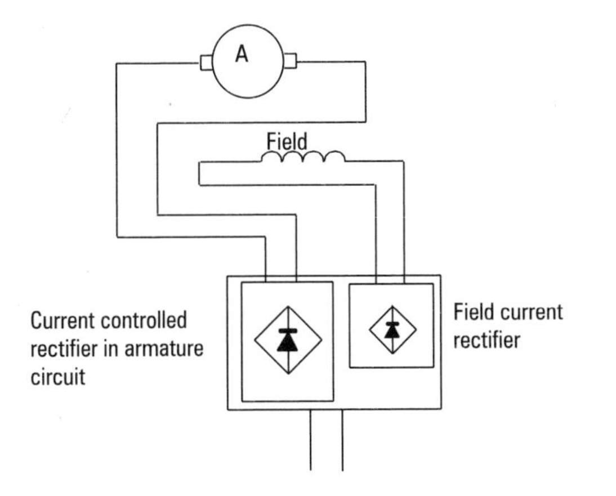

Figure 2.33 An a.c. supply connected to a d.c. shunt motor and rectifier units.

A controlled rectifier, or a simple rectifier followed by a chopper, would be used to supply the armature circuit and an independent control would be used for the field. If it is important to keep the supply power factor to a minimum, then a simple rectifier and chopper would be used, but this is more expensive than the controlled rectifier. On large machines, over 100 kW, choppers are seldom used and controlled rectifiers are far more likely.

Figure 2.34 shows an arrangement where the field winding of a shunt connected motor is controlled independently of the armature circuit. This ensures that the speed/torque characteristics can be maintained. The controllers shown in the diagram are supplied with three-phase to give a high d.c. output. Single-phase supplies could be used where necessary. The armature circuit is designed so that the two converters ensure a constant d.c. current from all of the a.c. waveforms.

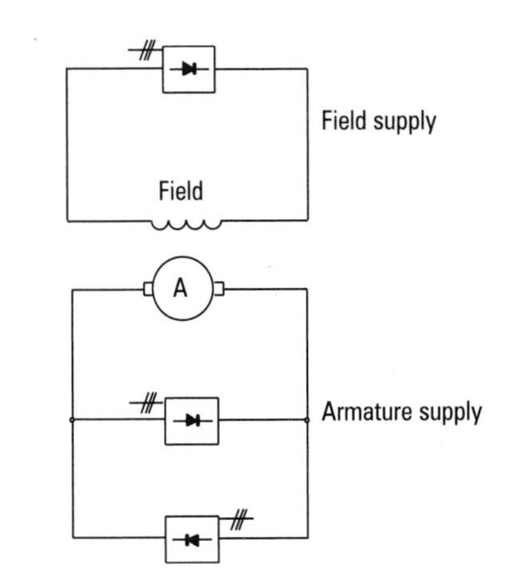

Figure 2.34

Ward–Leonard method of speed control

As the Ward–Leonard system uses a combination of d.c. generators and d.c. motors to produce the speed control, this has been covered at the end of Chapter 3.

Exercises

1. Explain why
   (a) the iron core of the armature of a d.c. machine is "laminated".
   (b) the brush that makes the connection is usually made of a carbon compound and not solid copper.

2. (a) With the assistance of at least two diagrams explain how a compound motor may be connected as
   i) short shunt
   ii) long shunt
   (b) Explain the dangers of weakening the strength of the series winding on a differentially compounded motor.

3. (a) Explain how a variable resistor connected in series with the shunt winding of a d.c. shunt connected motor can vary the armature speed.
   (b) A shunt motor is connected to a 250 V d.c. supply and has a field resistance of 50 Ω. Calculate
   i) the armature current when the motor takes a current of 75 A from the supply.
   ii) the armature resistance if it works on an induced voltage of 235 V.

4. (a) Explain why it is necessary to provide a starter for large d.c. motors.
   (b) A 25 kW, 230 V d.c. shunt motor has an efficiency of 75%. The shunt field has a resistance of 120 Ω and the armature resistance is 0.12 Ω. Assuming the voltage drop at the brushes is 2.5 V, determine the
   i) armature current
   ii) generated e.m.f.
   iii) total copper losses.

3

D.C. Generators

Complete the following to remind yourself of some important facts that you should remember from the previous chapter.

The magnetic field in most motors is produced by ________.

In a magnetic circuit for a two pole motor, the magnetic flux flows through ______ pole, round the ________, then through the ________ pole and across to the ________ pole through the steel ________ of the ________ to complete the ________ circuit.

To control the speed of a series motor __________ resistors can be connected across either the ________ or ________ ____________.

Variable resistances are connected into the __________ circuit of a shunt wound motor to __________ the ______ ___________.

To reverse the direction of rotation of a compound motor either change the ______________ connections or change ________ ______ winding connections.

On large d.c. motors ______ reaction problems can be overcome by fitting ________ between the ______ poles of the __________ windings.

The commutator is like a ________ switch which ______ the direction of the __________ in the ________ coils of the motor.

On completion of this chapter you should be able to:

- describe the construction of d.c. generators
- relate the type of generator best suited to a particular application
- describe the characteristics of given types of d.c. generator
- describe the Ward–Leonard system of speed control

D.C. generators and motors are basically the same machine. Most direct current machines will work as either motors or generators as the construction is basically the same. As with d.c. motors there are several different methods of connecting the windings and the generator names are taken from the connections.

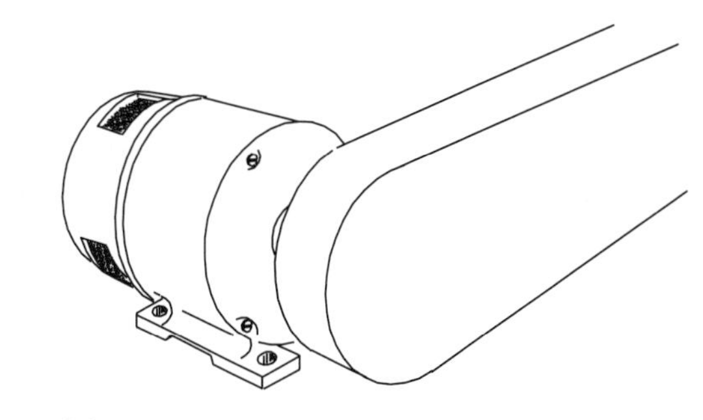

Figure 3.1

Construction

Materials used in construction

Case	iron, to complete the magnetic circuit
Armature core	magnetic steel laminations, to reduce hysteresis and eddy currents
Commutator	copper segments
Brushes	carbon composition, negative temperature/resistance coefficient
Field windings	copper wire formed into coils
Field poles	cast iron ground into shape

Construction

The basic construction of d.c. generators is the same as d.c. motors. Figure 3.2 shows an exploded view of a typical two pole generator. The field windings are preformed to the shape of the pole they are to fit. The pole is bolted to the case of the machine so that it forms part of the magnetic circuit. On generators the pole shoe is made of solid iron so that some residual magnetism is held when the machine is switched off. The armature and brush arrangements are the same as for motors.

Construction

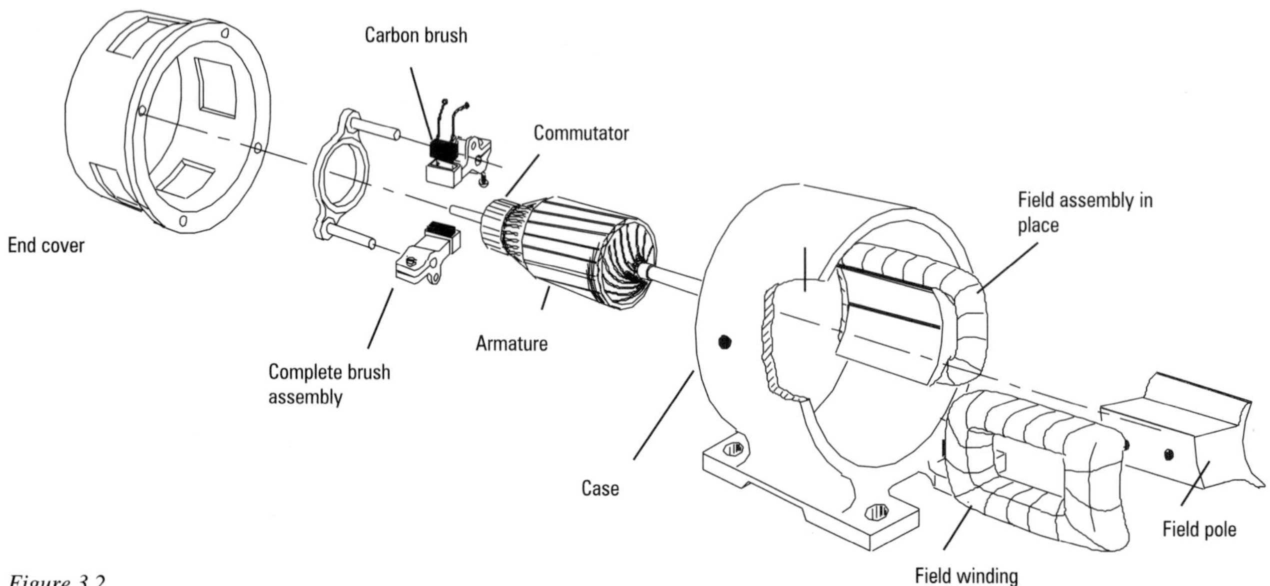

Figure 3.2

The theory of the generator is that the armature is rotated by an external prime mover. This may be in the form of an engine or a method of using natural resources such as the wind. As the armature conductors rotate, they move through the magnetic field produced by the field windings. This magnetic field will vary in strength depending on where it is supplied from. Separately excited machines, which take their field supply from an external d.c. source, have a strong magnetic field from stationary. Self excited generators have to rely on the residual magnetism left in the poles to get the first e.m.f.s generated.

When the armature conductors move through the magnetic field an emf is induced into the armature windings,

$$E \propto \Phi N$$

It can be seen from this that the strength of the magnetic flux has a direct relationship to the generated e.m.f.

The applications of the generators vary with the characteristics for the different types. Some are used far more than others. The series generator, for example, has very few applications and is therefore seldom found in use these days.

In Chapter 2 "Armature reaction" and "Commutation" were explained in some detail. Although they have not been covered again in this chapter, the theory is equally as important. Interpoles are used in most large generators and they are connected in series with the armature the same as in motors. The use of carbon compositions for brushes also applies to generators as a way of reducing the effects of commutation.

Separately excited generators

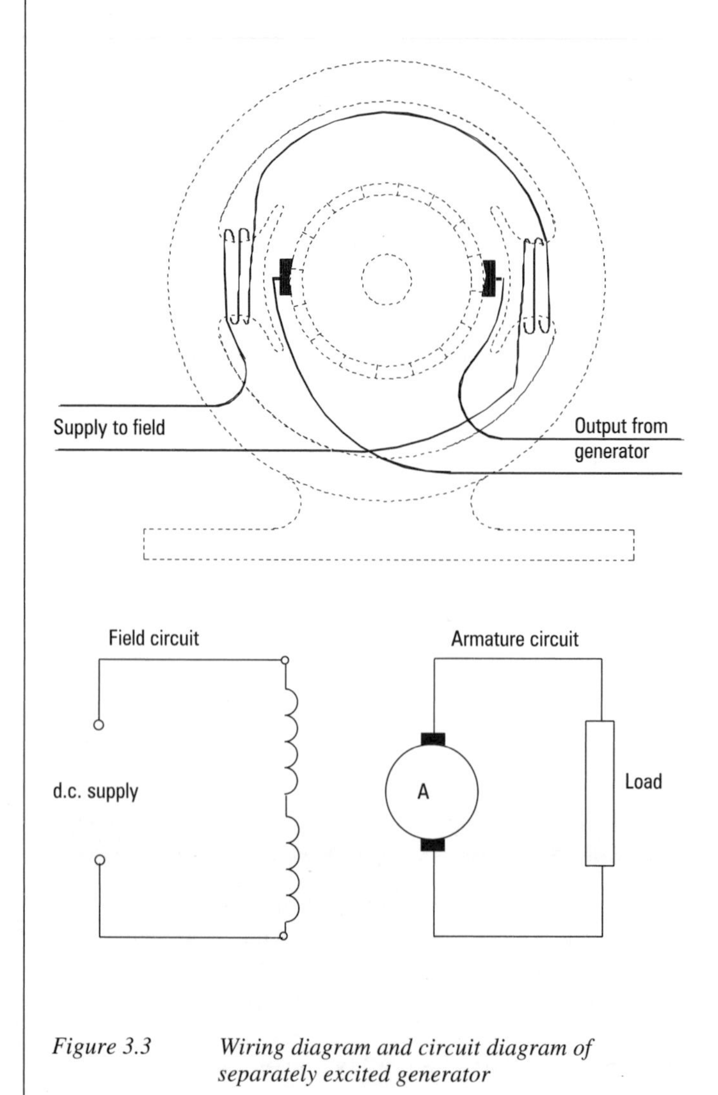

Figure 3.3 *Wiring diagram and circuit diagram of separately excited generator*

As the name implies, the field windings have an external d.c. power supply connected to them to provide the magnetic flux. This supply is often a battery unit which may have a variable resistor incorporated into the circuit to vary the output. The generated output of the machine is taken directly from across the armature.

The output of this type of generator can be fairly constant but the voltage tends to drop off slightly as the load current is increased. This is due to the effect of the load current and armature resistance causing a voltage drop in the armature. The characteristics for this type of machine are shown in Figure 3.4.

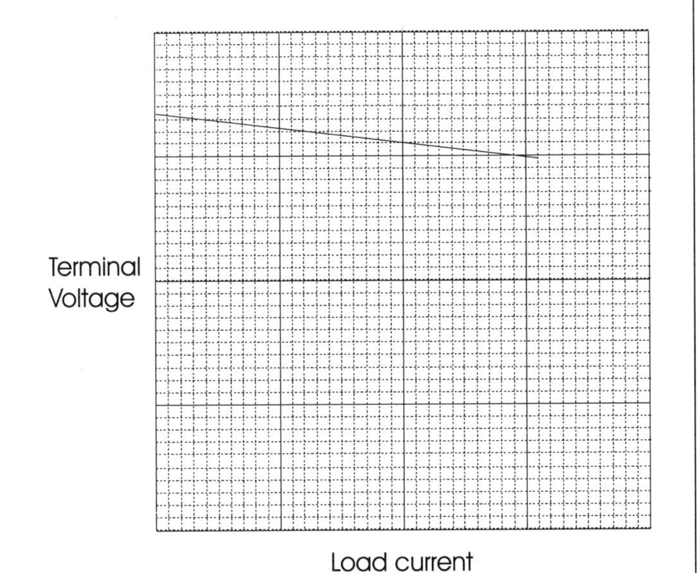

Figure 3.4 *Load characteristics of a separately excited generator*

Separately excited generator summary

Windings
Separate and connected in parallel

External supply
d.c. to field windings

Output voltage
Drops off as the load increases.

Regulation
Output voltage can be regulated by varying the field voltage.

Reversing output polarity
Change either field or armature but **NOT** both.

Try this

Before continuing list below some applications for d.c. generators.

Self excited generators

Unlike the separately excited generators there is no external supply in this type to power the electromagnets that produce the field. The field windings receive their supply from that produced by the armature. This means that there must be some residual magnetism in the poles for a start so that an initial voltage can be created. Once the armature is turning and current starts to flow in the circuits the magnetic field becomes stronger and the supply voltage increases. If there is no residual magnetism to start with no voltage can be generated. Similarly if the machine is started so that the residual magnetism is weakened instead of strengthened, it will not excite and therefore not generate.

Examples of self excited generators are:

- the series wound generator
- the shunt wound generator
- the compound wound generator

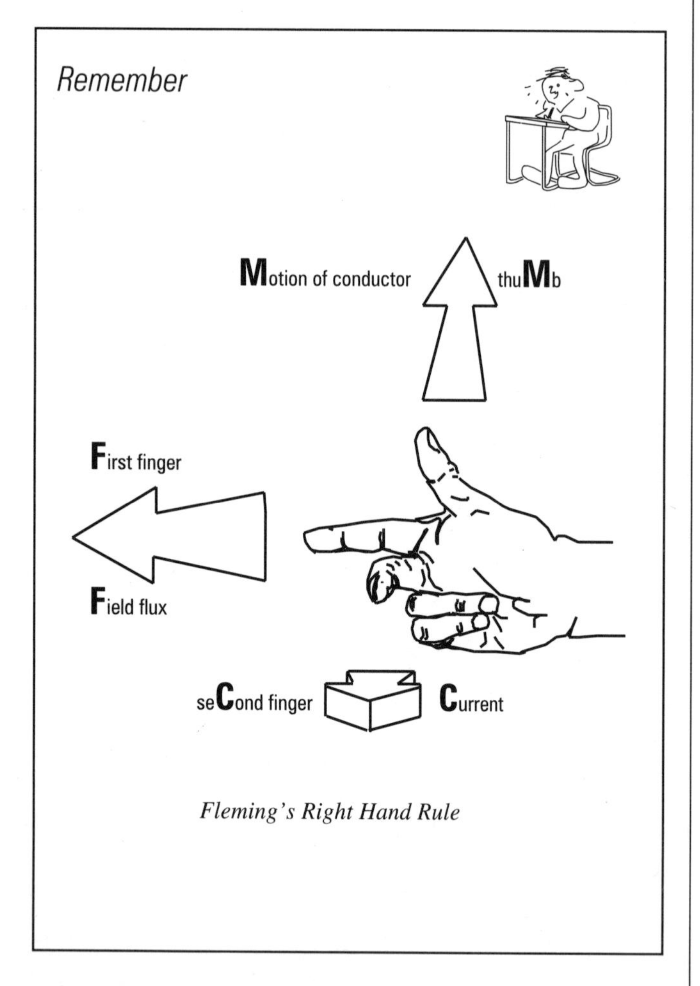

Fleming's Right Hand Rule

Series wound generators

In a series connected generator the current flowing in the field windings is the same as that in the armature. The field winding is therefore made of a comparatively few turns of very thick wire.

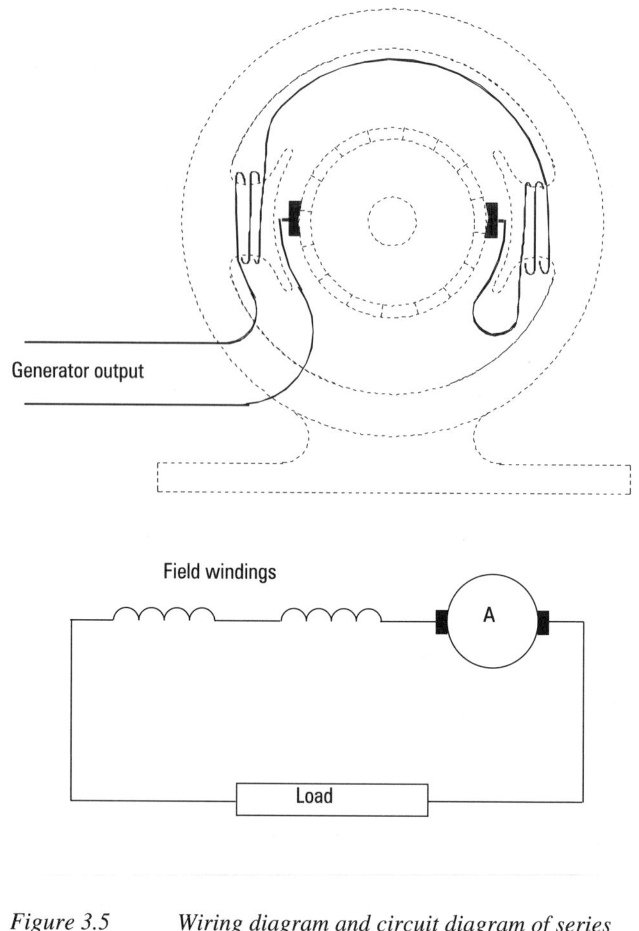

Figure 3.5 *Wiring diagram and circuit diagram of series connected generator*

When the current being supplied by the generator is small, the current flowing in the field is small and the magnetic field is weak. This results in a low voltage being generated. As the load current is increased so the field strength is increased and a higher voltage is generated. This effect can be seen in the characteristics shown in Figure 3.6.

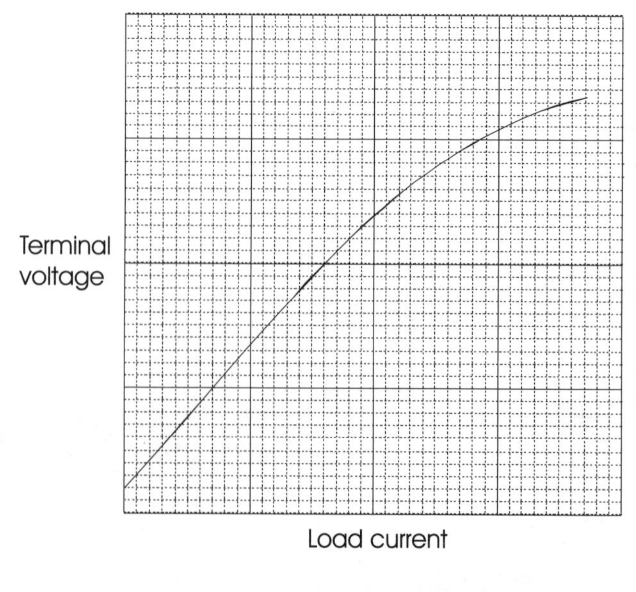

Figure 3.6 *Load characteristics for a series connected generator*

On no load there is a small voltage generated due to the residual magnetism in the poles. The use of this type of generator is very limited but in the past they have been used to boost the voltage to overcome voltage drop in long cable runs.

Series wound generator summary

Windings
Armature and field connected in series

External supply
None

Output voltage
This increases as the load current increases.

Reversing output polarity
Change either field or armature but **NOT** both.

Example

Consider a series wound d.c. generator that supplies a current of 100 A and has:

- an armature resistance of 0.1 ohm
- a field resistance of 0.2 ohms
- a generated e.m.f. of 260 V
- a brush contact voltage drop of 5 V

The terminal voltage has to be calculated.

The circuit diagram can be seen in Figure 3.7.

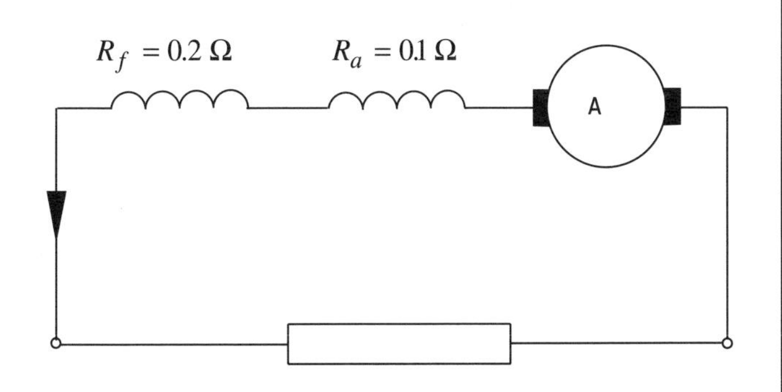

Figure 3.7 Circuit diagram for example

The terminal voltage

$$U = E - I_a R_a - I_s R_f - U_b$$

Where:

E	=	generated e.m.f
I_a	=	armature current
R_a	=	armature resistance
I_s	=	series field current
R_f	=	series field resistance
U_b	=	voltage drop due to brush contact

$$U = 260 - (0.1 \times 100) - (0.2 \times 100) - 5$$
$$= 260 - 10 - 20 - 5$$
$$= 225\ \text{V}$$

Try this

A series wound d.c. generator supplies 30 A at 200 V.

The generator has:

an armature resistance of 0.35 ohms
a field resistance of 0.5 ohms
and a brush contact voltage drop of 2 V.

Calculate the generated e.m.f.

The shunt wound generator

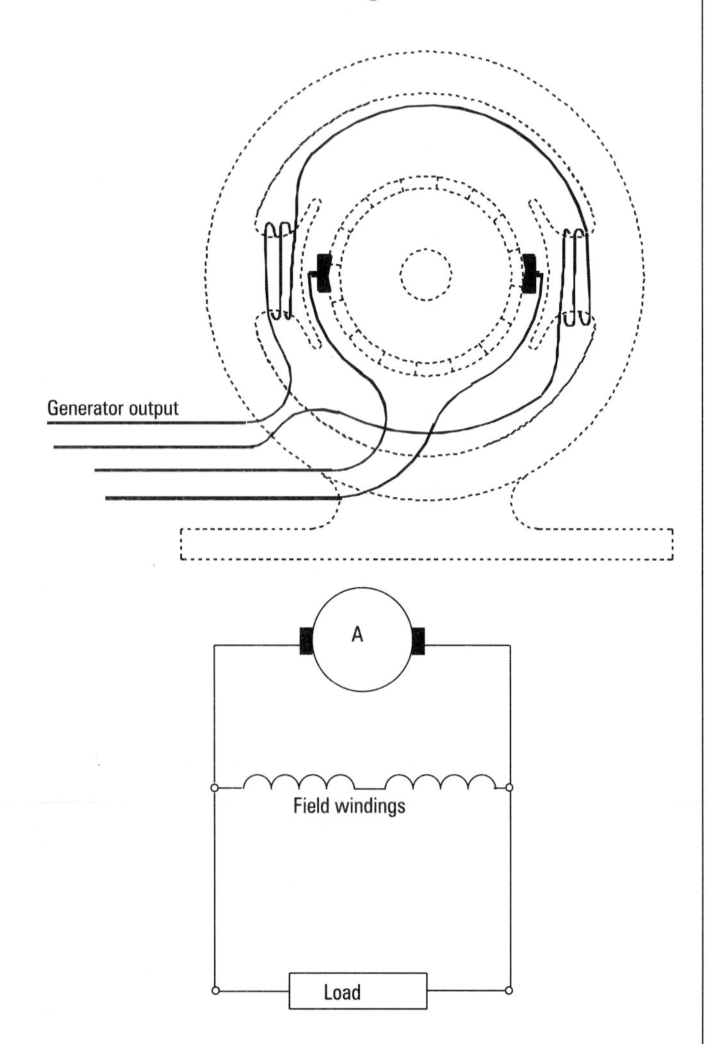

Figure 3.8 *Wiring diagram and circuit diagram of series connected generator*

The field and armature windings are connected in parallel (shunt) with each other (Figure 3.8). This means that the strength of the magnetic field is determined by the resistance of the field windings and the generated voltage. In practice the field winding is made up of coils with a large number of turns with comparatively thin wire. The power absorbed by the field is usually only about 2% of the generator's rated output.

The terminal voltage of this type of generator is fairly constant in a short range of loads. If the load is increased beyond this a voltage drop starts to appear. At first this voltage drop is due to the resistance of the armature windings but as a drop in terminal voltage also affects the field windings, a reduced magnetic field results. This in turn means that an even smaller voltage is generated so an even smaller terminal voltage is produced. If the load current is significantly increased the terminal voltage continues to drop of, as the characteristics show in Figure 3.9.

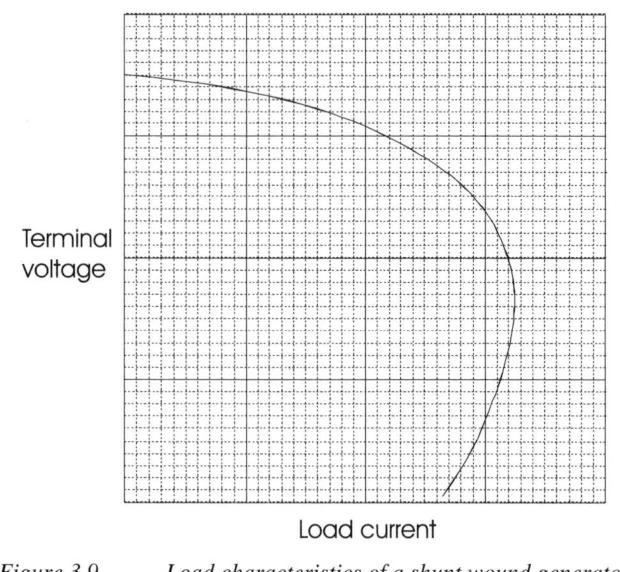

Figure 3.9 *Load characteristics of a shunt wound generator*

The terminal voltage can be controlled to some extent using a field connected regulator. By varying the resistance of the field circuit the current can be controlled and hence the output voltage (Figure 3.10).

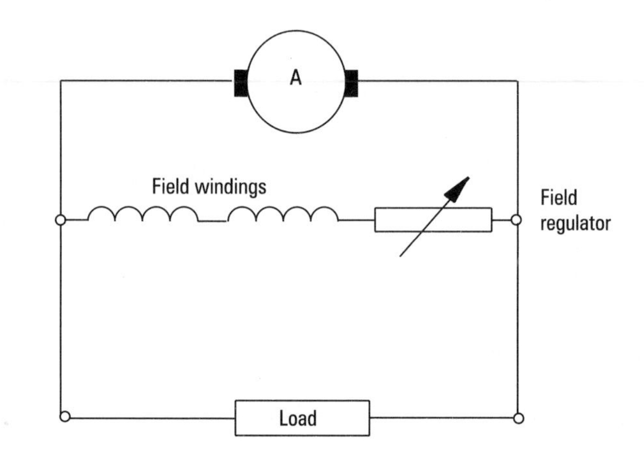

Figure 3.10 *Circuit diagram showing a field regulator in a shunt wound generator circuit*

Try this

Describe a typical application for a shunt wound generator.

Example

Consider a d.c. shunt connected generator that delivers a current of 85 A at 200 V. If it has:

an armature resistance of 0.15 ohms
a field winding resistance of 50 ohms
and a brush contact voltage drop of 1.8 V

the armature current (i), and the generated e.m.f. (ii) can be calculated.

The circuit diagram for this circuit is shown in Figure 3.11.

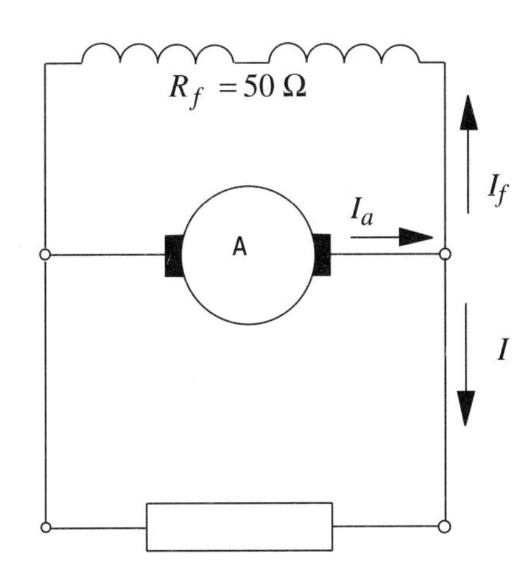

Figure 3.11 Circuit diagram for example

(i) the armature current is the total of the supply current and the current in the field winding.

Current in the field winding

$$I_f = \frac{U}{R_f}$$

$$= \frac{200}{50}$$

$$= 4\ \text{A}$$

Current in armature

$$I_a = I_f + \text{load current}$$
$$= 4 + 85$$
$$= 89\ \text{A}$$

(ii) The supply voltage = generated voltage – voltage drop in armature – volt drop due to brush contact

$$U = E - I_a R_a - U_b$$

Transposed

$$E = U + I_a R_a + U_b$$
$$= 200 + (89 \times 0.15) + 1.8$$
$$= 215.15\ \text{V}$$

Shunt wound generator summary

Windings

Armature and field windings in parallel

External supply

None

Output voltage

Constant over short range but drops off as load increases

Regulation

Variable resistor in series with field winding.

Reversing output polarity

Change armature or field winding connections **NOT** both.

Try this

A d.c. shunt generator supplies a load of 50 A at 220 V and has -

an armature resistance of 0.2 ohms
a field resistance of 115 ohms
and a brush contact voltage drop of 2 V

Calculate the generated e.m.f.

Try this

A d.c. shunt generator delivers 45 A at 250 V.

It has:

a field resistance of 100 ohms
a field current of 2.5 A
the armature generates 258 V
and the brush contact voltage drop is 2 V

Calculate
(i) the armature current
(ii) the armature resistance

Compound wound generator

The compound wound generator, as with the similarly wound motor, has both shunt and series connected windings. There are two ways of connecting the shunt windings as can be seen from Figures 3.12a and 3.12b. A short shunt connected machine takes the shunt winding directly across the armature, whereas a long shunt type takes it across the output terminals. In practice this makes very little difference as the series turns are very few and made of comparatively thick wire, unlike the shunt windings that consist of a large number of turns of comparatively thin wire. In either case the two sets of coils are usually connected so that their ampere-turns assist each other.

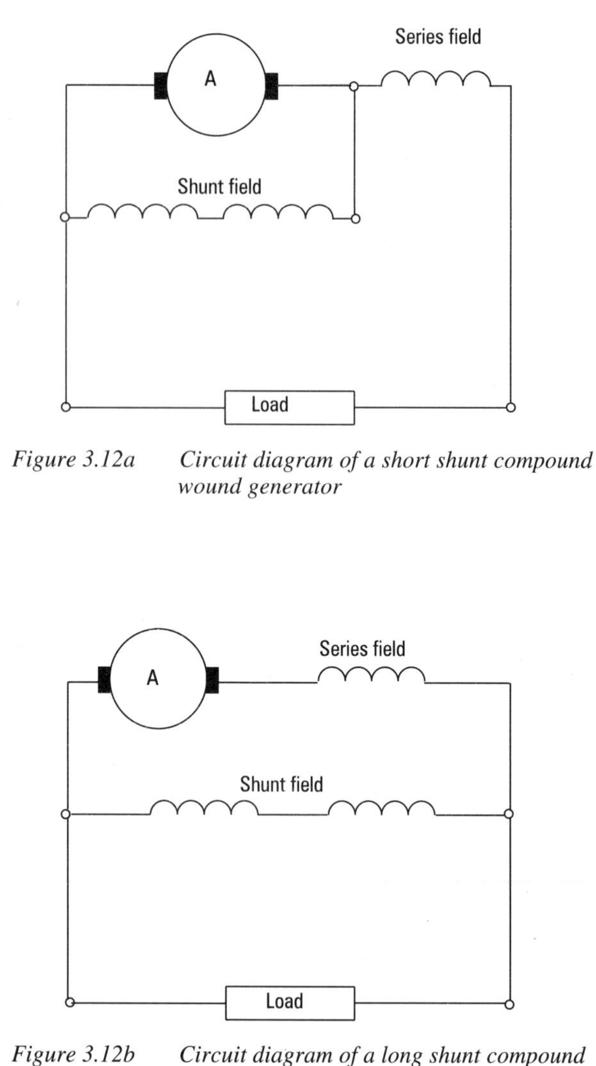

Figure 3.12a *Circuit diagram of a short shunt compound wound generator*

Figure 3.12b *Circuit diagram of a long shunt compound wound generator*

A generator that is designed and connected so that the currents set up in both the series and shunt windings assist each other is said to be "cumulatively compounded". If the shunt winding is the dominant one, and the series has little effect, this is described as being under-compounded. (Figure 3.13 curve A). To level out the curve more series ampere-turns need to be introduced, as curve B in Figure 3.13. In some instances generators are "over-compounded" to meet particular applications. The characteristic for these are shown in Figure 3.13 (curve C).

Generators that are designed and connected so that the field currents oppose each other are called "differentially compounded" and have characteristics as shown in Figure 3.13 (curve D).

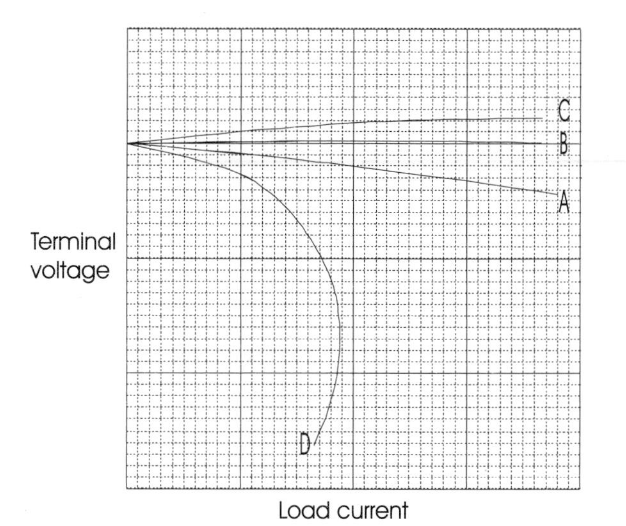

Figure 3.13 *Characteristics of compound generators showing:*
A – under compounded
B – level compound
C – over compound
D – differentially compounded

Compound wound generator summary

Windings
Armature and part field windings in series, separate field in parallel

External supply
None

Output voltage
Dependent on the method of compounding used.

Reversing output polarity
Changing either the armature or both sets of field winding connections but **NOT** both.

The Ward–Leonard System of Speed Control

Although this relates to the speed control of d.c. shunt motors, until the theory of the shunt wound generator was covered it was not practical to look at this system. Figure 3.14 shows how the armature of a d.c. shunt motor that has to have the speed controlled, is connected to the output of the generator. The field of the motor is separately supplied from a constant d.c. source. The d.c. generator is driven by a direct shaft coupling to a drive motor, which may be d.c. or, a.c. single or three-phase depending on the loads to be driven and the supplies available.

By controlling the strength of the field in the generator the output voltage can be varied from zero to maximum. If a change-over switch is used to reverse the field winding polarity, then the generator output can be varied from zero to maximum in both directions. In Chapter Two it was seen that adjusting the voltage across the armature is one of the methods of varying the motor speed. Doing this by the output of a generator rather than by variable resistors reduces the losses and produces greater efficiency at low speeds. It also means that a single controller in the generator field circuit can give a complete range of output speeds from the motor.

Although Figure 3.14 shows a field regulator in the generator circuit, this can be replaced on modern equipment, with sophisticated electronics. This modern type of controller can be designed so that the output speed is monitored and the information is fed back so that automatic adjustments are made to maintain constant speeds under varying loads.

The disadvantage of this system of speed control compared to electronic methods is the comparatively high losses and low efficiency of the machines.

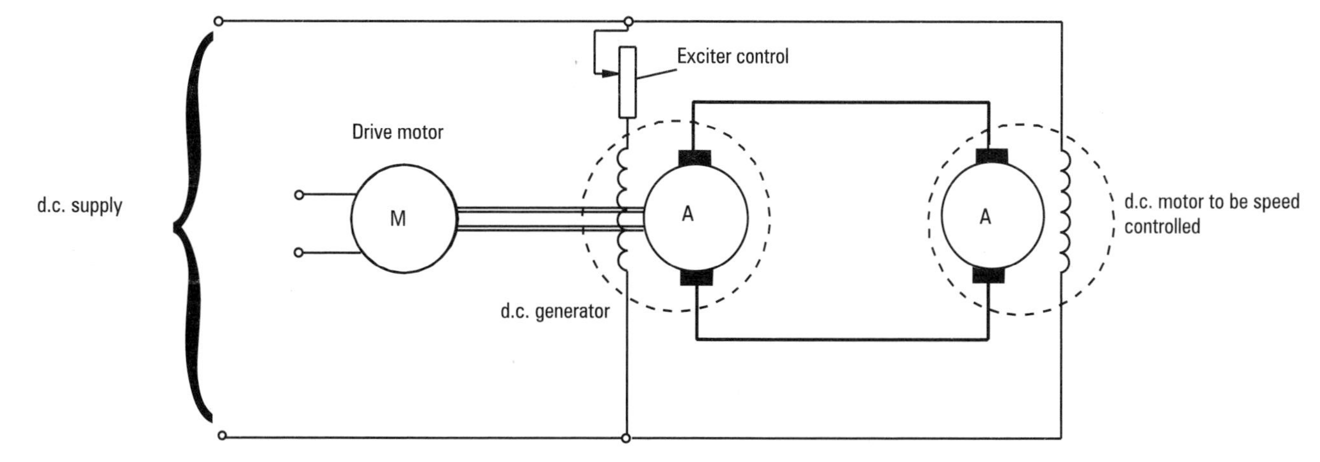

Figure 3.14

Exercises

1. A shunt wound generator supplies a voltage of 150 V at a constant load of 150 A. If it has an armature resistance of 0.03 ohms, a field resistance of 50 ohms, and a voltage drop across the brush gear of 2.15 V, calculate the
   (a) field current
   (b) armature current
   (c) generated e.m.f.
   (d) maximum voltage that can be expected at the end of a cable run with a total resistance of 0.05 ohms when it carries 150 A.

2. (a) Describe with the aid of diagrams the construction of a typical field pole and winding from a d.c. shunt connected generator.
   (b) A d.c. shunt wound generator has an armature resistance of 0.18 ohms and a field resistance of 110 ohms. Calculate the currents in
   (i) the field winding and the
   (ii) armature
   when supplying 150 A at 450 V.

3. (a) The characteristics of a d.c. shunt wound generator are:

Terminal voltage	(V)	300	290	280	265	250	150
Load current	(A)	0	25	50	75	100	125

Draw a curve of the above results and suggest what the terminal voltage would be when supplying a constant load of 85 A.

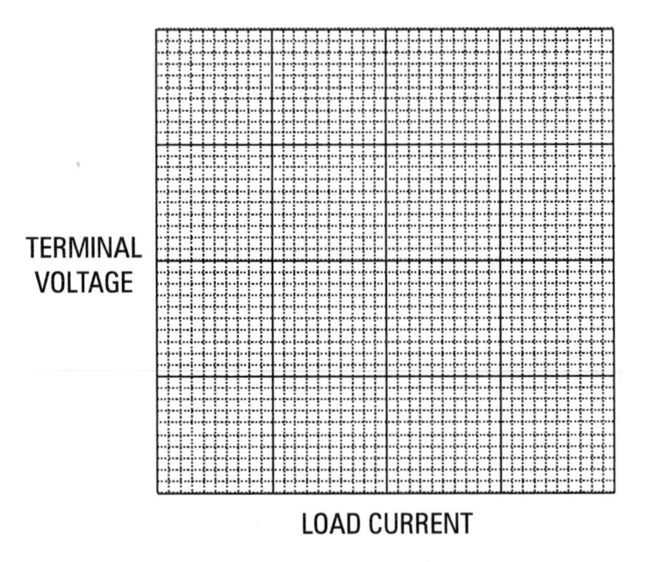

(b) If the generator in (a) has an armature resistance of 0.15 ohms and current of 50 A calculate the generated e.m.f. Assume that the voltage drop due to brush resistance is 2.5 V.

4. A shunt wound d.c. generator, driven at constant speed, supplies a load of 21 kW at 150 V. The resistance of the armature is 0.021 W and there is a total voltage drop of 2.25 V in the brush gear. If the resistance of the shunt field is 62 W, calculate:
   (a) the load current
   (b) the generator field current
   (c) the generator armature current
   (d) the generated e.m.f.

4

A.C. Motors

Complete the following to remind yourself of some important facts that you should remember from the previous chapter.

The generator armature is rotated by an external ________ ______________.

Separately excited generators obtain their field supply from an _________ d.c. ___________, and they have a ____ magnetic field from __________. Self excited generators have to rely on the ________ ________ left in the poles to get the first _________ generated.

The output voltage of a separately excited generator tends to drop off as the ________ ___________ ________ .

A self excited d.c. generator will not ________ an output voltage if there is no ______________ magnetism in the field ______________.

The output voltage of a shunt-wound generator is ______ over a ________ range of loads.

When a compound wound generator is wound with both the ______ and ______ windings assisting each other, it is ____________ compounded, and if these fields are wound to oppose each other it is _____________ __________.

In a Ward–Leonard System the motor driving the generator may be ________, or ______ _______ or ______ phase.

On completion of this chapter you should be able to:

- ◆ describe the construction of an a.c. cage induction motor
- ◆ describe the production of a rotating magnetic field from a three-phase supply
- ◆ describe the production of a rotating magnetic field from a single-phase supply
- ◆ calculate the speed of a.c. motors from supply frequency and number of poles
- ◆ identify the different methods of starting a.c. motors
- ◆ recognise the need for adequate overcurrent protection
- ◆ describe various control circuits used for a.c. motors
- ◆ describe the method of starting wound rotor induction motors
- ◆ recognise the need and methods for using solid-state soft start controller starters
- ◆ describe the construction and operation of pole change motors

When looking at d.c. motors it was necessary to revise on basic magnetic theory as rotation was created by producing a force on a conductor in a magnetic field. Although a.c. motors still work using the principles of magnetism, they also rely on the principle of electromagnetic induction.

In Chapter 1, the basic production of an a.c. wave form was used when looking at the development of the generator. This wave form is of course single-phase and is very limited as a supply for a.c. motors. A three-phase supply on the other hand, is ideal and can be used to great effect.

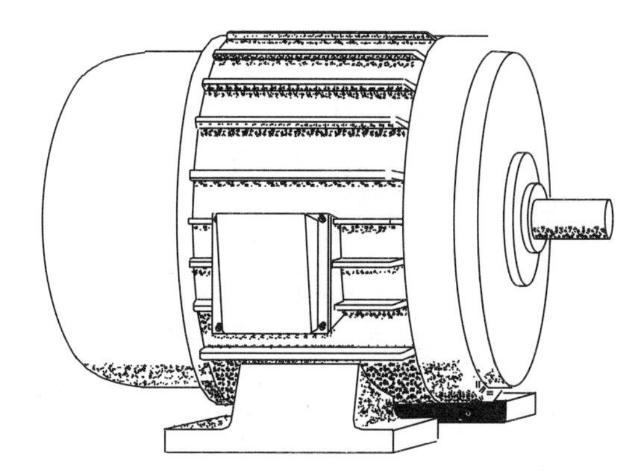

Figure 4.1

Construction

The construction of an a.c. motor is usually completely different to that of a d.c. The exception is the universal motor which will be dealt with later.

Basically an a.c. motor consists of, field windings (stator) which create a rotating magnetic field, and a rotor which reacts to the field.

Construction

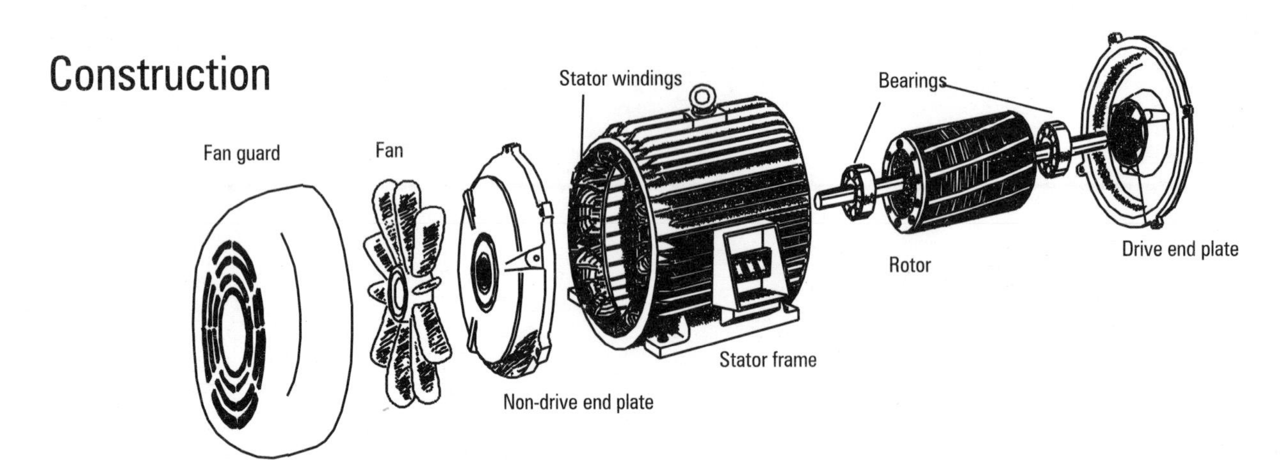

Figure 4.2 *a.c. motor construction*

The field windings are designed to use the three-phase supply consisting of the three wave forms displaced by 120°, as shown in Figure 4.3.

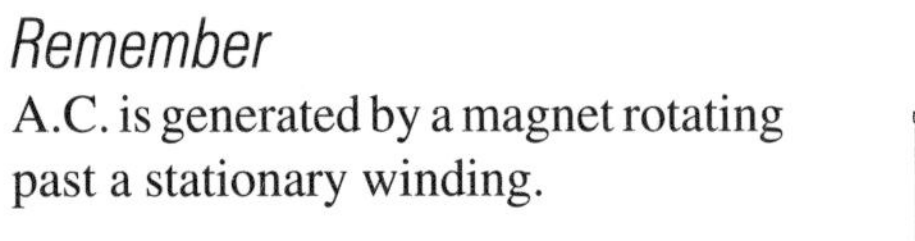

Remember
A.C. is generated by a magnet rotating past a stationary winding.

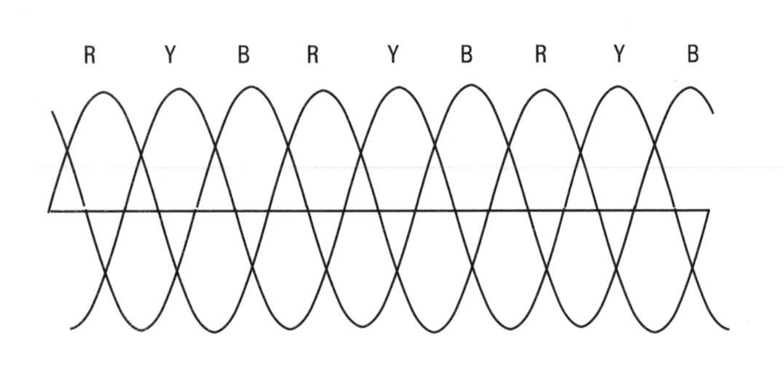

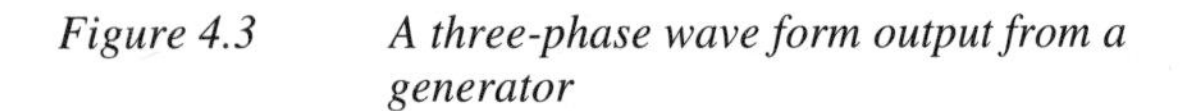

Figure 4.3 *A three-phase wave form output from a generator*

The stator

The field windings of a three-phase are arranged so that there are three sets each placed 120° apart. As each winding is connected to a separate phase, the peak current in each coil occurs at different times. In Figure 4.4 it can be seen that for the instant indicated the currents in the windings are as shown on the ammeters.

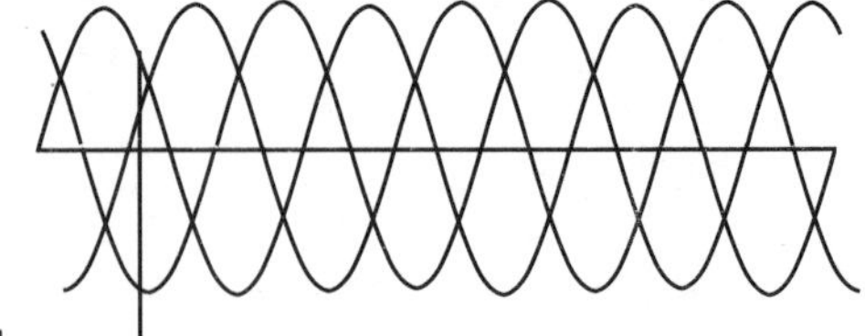

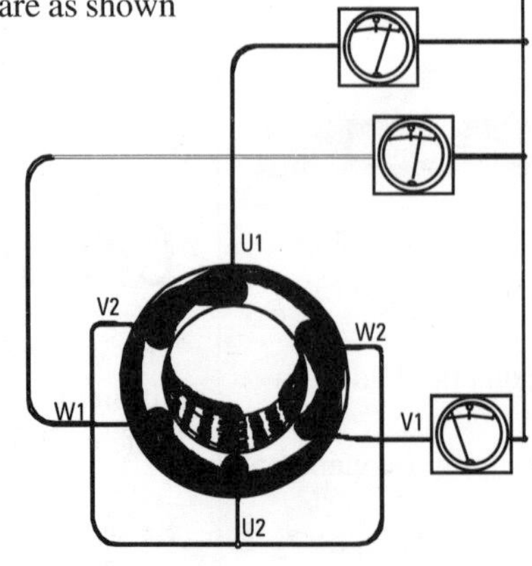

Figure 4.4 *Simplified field winding assembly*

The field windings are designed to reflect the potential movement in the wave forms. A comparison can be drawn between the waves of a three-phase wave form and the waves of a river. A stick placed in a river is taken along on the waves from one place to another. In a three-phase wave pattern the direction of movement follows in the sequence red, yellow, blue. A magnetic conductor (stick) introduced onto the first wave (red) is carried up to the peak of the wave form and down the other side until it is picked up by the yellow wave and carried on. This will continue all the time there is a series of wave forms to carry it. If instead of the wave forms being on a straight axis they are incorporated into the field winding of a motor, they are then circular and the wave forms will have the effect of going round in circles. This effect can be seen by placing a compass into the centre of a motor field winding when it is connected to a three-phase supply. The magnetic pole of the compass needle follows the rotating magnetic field set up in the windings.

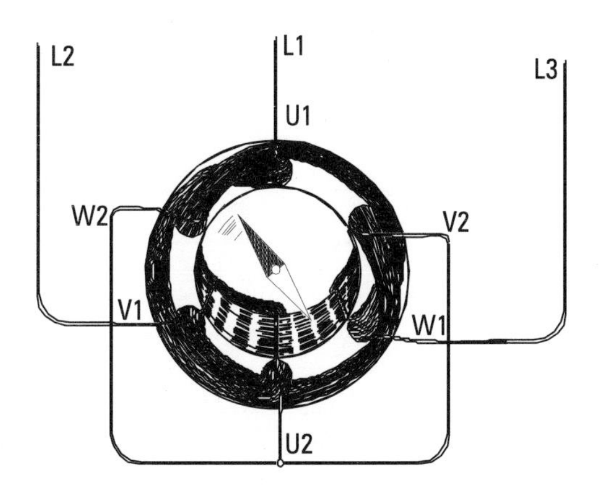

Figure 4.5 *Only one set of poles for each phase are shown for clarity*

To reverse the direction of rotation of the compass needle, the phase sequence of red, yellow, blue must be reversed. This can be achieved by changing over any two of the phases. So the sequence

red ⇒ yellow ⇒ blue ⇒ red ⇒ yellow ⇒ blue becomes
yellow ⇐ red ⇐ blue ⇐ yellow ⇐ red ⇐ blue

when the red and yellow are changed over. Now to read the sequence red, yellow, blue, they have to be read from right to left instead of left to right.

The rotor

Unlike d.c. motors, a.c. induction motors do not have electrical connections to the rotating parts of the machine. The rotor of an induction motor can be likened to the secondary winding of a double-wound transformer. The primary winding being the field winding in the stator.

The rotor consists of a number of copper or aluminium conductors embedded in a laminated iron cylinder. Figures 4.6 and 4.7 show a typical construction.

Figure 4.6 *Rotor with laminations*

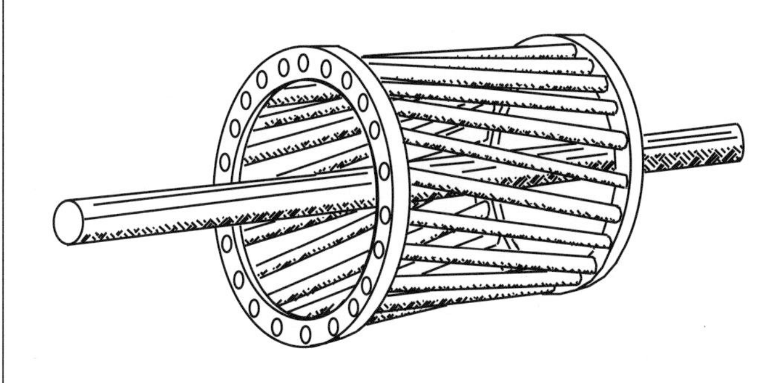

Figure 4.7 *Rotor with laminations removed*

It is this construction of the rotor that has given this type of motor the name "cage rotor".

When the rotor is fitted into the field winding the gap between the two sets of laminations is very small. This keeps the reluctance of the magnetic circuit and losses to a minimum. When current flows in the field windings, its alternating movement induces a magnetic flux in the rotor bars. As these bars are made of large conductors which are shorted out by rings at each end of the rotor, large currents flow. These currents produce their own magnetic fields which react with the magnetic flux set up in the field windings. As the magnetic flux rotates in the field windings, the rotor magnetic field tries to follow it. This is how the rotation is created in an a.c. induction motor. There are no fixed magnetic poles like on d.c. machines, the magnetic field produced by the supply is continually rotating. The magnetic field in the rotor, resulting from the induced current, is always trying to catch up with the rotating field in the stator.

Speed

The speed of an a.c. induction motor is directly related to the frequency of the supply and the number of poles in the motor. A motor with a stator which has one pair of poles per phase, completes one revolution of 360° in one complete cycle. A four pole machine completes one revolution in every two cycles.

Assuming a two pole motor is working on 50 Hz, its ideal speed is

$$\text{speed} = \frac{\text{frequency}}{\text{no. of pairs of poles}} \text{ in revs per sec}$$

$$n_s = \frac{f}{p} \text{ revs per sec}$$

$$n_s = \frac{50}{1}$$

$$= 50 \text{ revs per sec}$$

$$\text{Or } N_s = 50 \times 60$$
$$= 3000 \text{ revs per min}$$

where

n_s = speed in revolutions per second (rev/sec)
N_s = speed in revolutions per minute (rev/min)
f = frequency of the supply in Hertz (Hz)
p = number of pair of poles per phase

This is an ideal situation and assumes that the rotor runs at the speed of the rotating magnetic field in the stator. This is known as synchronous speed. Under normal conditions the rotor speed will be less than synchronous speed due to what is known as "slip". A motor that does not run at synchronous speed is sometimes referred to as an asynchronous motor.

If the rotor was to run at synchronous speed the rotor bars would be moving parallel to the rotating field in the stator. This would mean that the current would appear to be stationary relative to the rotor. Under these conditions no e.m.f. would be induced into the rotor and therefore there would be no magnetic field and the rotor would develop no torque. As the rotor speed then slows down its relative speed with the rotating magnetic field increases and higher values of e.m.f. are induced. The torque is therefore increased and reaches that required by the load connected to the motor.

Slip can be calculated in two forms, either per unit slip sometimes also referred to as fractional slip, or percentage slip.

$$\text{per unit slip} = \frac{N_s - N_r}{N_s}$$

$$\text{percentage slip} = \frac{N_s - N_r}{N_s} \times 100\%$$

where

N_r = rotor speed (rev/min)

Table 4.1

Synchronous speed and standard rotor speeds at full load (50 Hz supply)

Poles	Synchronous speeds revs/min	Rotor speeds revs/min
2	3000	2900
4	1500	1440
6	1000	960
8	750	720
10	600	580
12	500	480
16	375	360

Try this

1. A four-pole motor has a rotor speed of 1444 revs/min when working on a supply frequency of 50 Hz. Calculate
   (a) the per unit slip
   (b) the percentage slip

2. A two-pole motor has a per unit slip of 0.03 on full load. Calculate the full load speed when it is working
   (a) on 50 Hz and
   (b) on 60 Hz

Remember

There must be relative movement between rotor conductors and the rotating magnetic field for torque to be produced.

Torque

When an induction motor is switched directly on to full line voltage the starting torque is approximately 130% that of the full load torque. The current at this time is in the region of 600% full load current. As the motor picks up speed the torque increases to its maximum of between 200% and 250% of the full load torque, at approximately 75% synchronous speed. The torque then falls away rapidly to meet the requirements to drive the load. Figure 4.8 shows typical characteristics for torque and current.

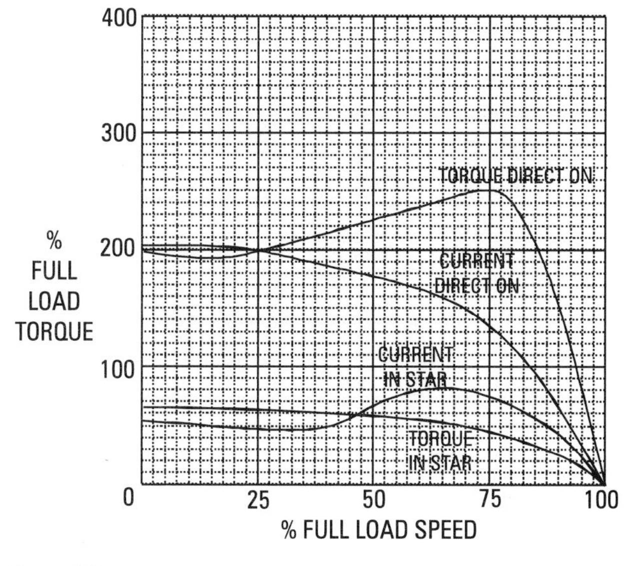

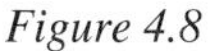
Figure 4.8

Improved starting torque and reduced starting current can be achieved by increasing the resistance of the rotor. For this special alloys have been developed for the rotor bars and shorting rings. A double cage rotor is often used for this purpose. In these the rotor is designed so that the motor operates with a high resistance circuit during starting, using the outer rotor bars, and low resistance under normal running conditions taking advantage of the low resistance inner bars (Figure 4.9).

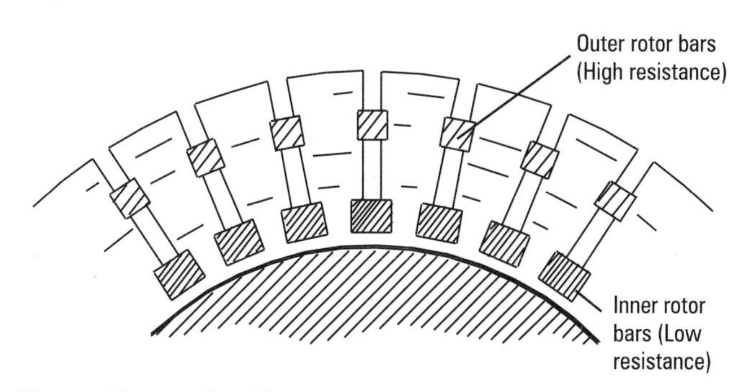

Figure 4.9 *Double cage rotor*

The result with the high resistance rotors is that although the starting torque is increased and the starting current is reduced there is an increase in slip and therefore a reduction in the efficiency of the motor.

A wound rotor slip-ring motor with external rotor resistances used for starting can be used. These give maximum torque at starting together with minimum resistance for the minimum slip and maximum efficiency, at full load speed. This type of motor is dealt with in more detail further on in this chapter.

Remember

Starting torque depends on the design of the rotor.

A double-cage rotor has a better starting torque than a single-cage rotor.

Efficiency

In an induction motor the input power is only supplied to the stator and any losses are related back to that. This can be shown as three parts.

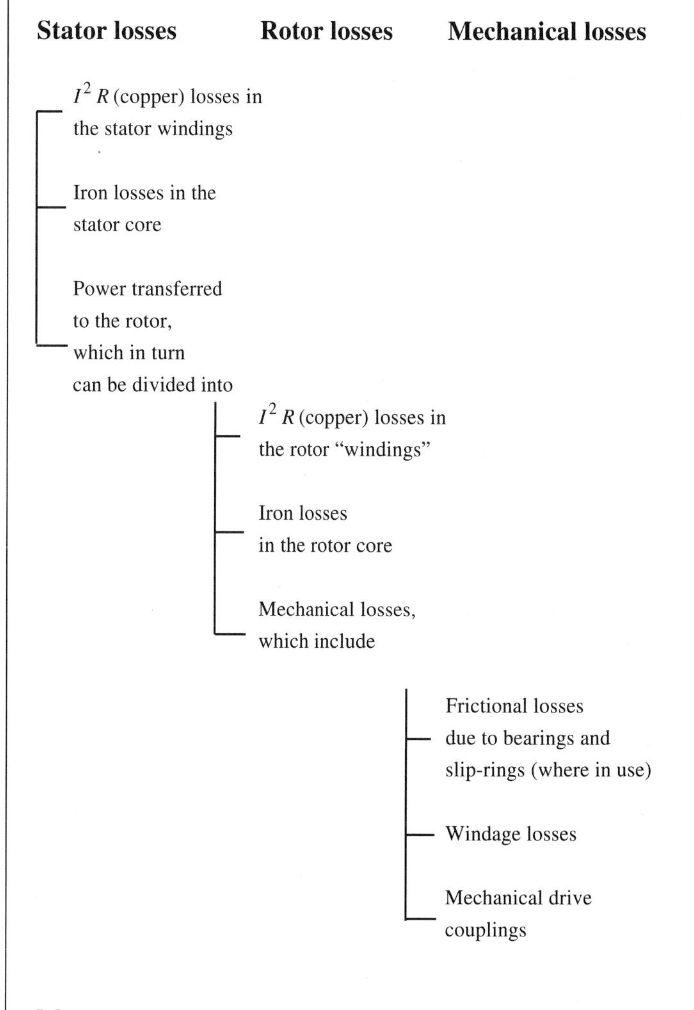

Motor starters

It can be seen from the characteristics in Figure 4.8 that the starting current of an induction motor when started directly across the supply, can be as high as 600% of the full load current. For this reason the use of direct-on-line starting is limited to motors of up to 4 kW by the Public Electricity Suppliers. In some areas larger motors are allowed depending on the capacity of mains supply.

It is a requirement of BS7671:1992 that all motors with a rating exceeding 0.37 kW must have protection against overload. There are also requirements with regard to the motor not automatically restarting after a voltage drop or supply failure. As all starting methods must include these safety measures they will be considered first.

Overload devices

Thermal type

The thermal type overload works on the principle of indirect heat on a bimetal strip. A coil of resistance wire is wound round a bimetal strip and connected in series with the motor. Under normal conditions the resistance wire does gain heat so the bimetal strip is not effected. When an overload develops and the motor takes more current, the resistance wire heats up and in turn passes the heat on to the bimetal strip. This bends and when it gets to a predetermined position, pushes an insulated strip that operates a switch which breaks the hold on circuit. The fault may be an overload that affects all three phases or an insulation breakdown affecting only one phase. The result in each case is a total disconnection of supply. Different heater/bimetal strip assemblies can be fitted for different loads. Some adjustment is usually fitted at the contact position for small variations in load. Figure 4.10 shows a typical type of overload detector.

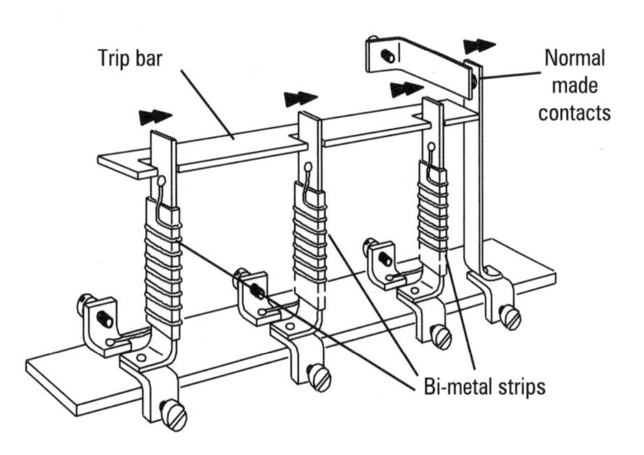

Figure 4.10

Magnetic type

A solenoid arrangement is connected into each load conductor of the motor being protected. Each solenoid is designed so that under normal conditions the magnetic flux created by the load current of the motor is not enough to lift the soft iron armature inside the coil. When the current increases due to excessive load or a fault, the flux increases and a force is exerted on the armature. At a predetermined level the armature extension pushes against an insulated bar which opens contacts and breaks the hold-on circuit. As unlike the thermal type, magnetism can operate very fast, a damping device has to be fitted to overcome the high starting currents. This usually consists of a dash pot filled with a light transformer oil and a piston fitted to the bottom of the armature. When there are high starting currents flowing the armature has to overcome the resistance of the piston in the oil as it tries to pull up through the solenoid. To allow for calibration there are two adjustments that can be made. First the dashpot can be screwed up and down, adjusting the length of travel the armature must make. Secondly there are holes in the piston that can be changed to adjust the amount of resistance the oil holds against the piston. Figure 4.11 shows an arrangement for this type of overload.

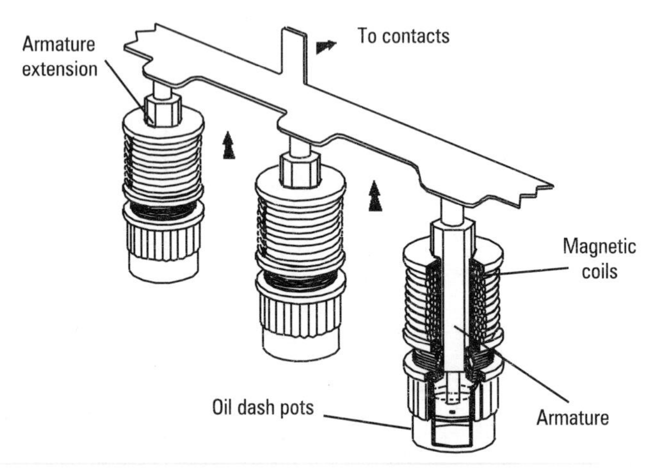

Figure 4.11 Magnetic overload assembly

Try this

Compare the advantages and disadvantages of both thermal and magnetic overload protection devices employed in induction motor starters.

Thermistor overload detection

A thermistor is a type of resistor that varies in resistance when there is a change in temperature. Depending on the type of thermistor in use the resistance will either increase or decrease with rise in temperature. The thermistors, which are usually not much larger than a match, are embedded directly into the motor windings. On a three-phase motor there would usually be three thermistors, one embedded in each winding. These are normally connected in series and wired back to a control unit in the starter. Figure 4.10 shows a typical arrangement.

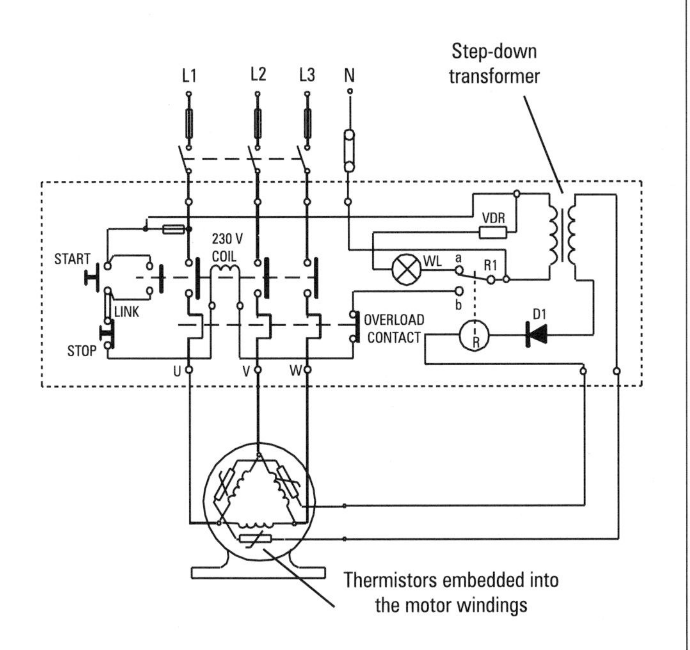

Figure 4.12 *A three-phase direct-on-line starter incorporating thermistor protection*

Key:

D_1	Diode
R	DC relay coil
R_1	Change-over contacts on relay
WL	Warning lamp
VDR	Voltage dropping resistor
+ t°	Positive temperature co-efficient type thermistor

When there is a change in temperature in any of the motor windings, due to an overload or fault, this is detected by a thermistor. The change in resistance that results is relayed back to the starter and when it reaches a predetermined level, the starter disconnects the supply to the motor.

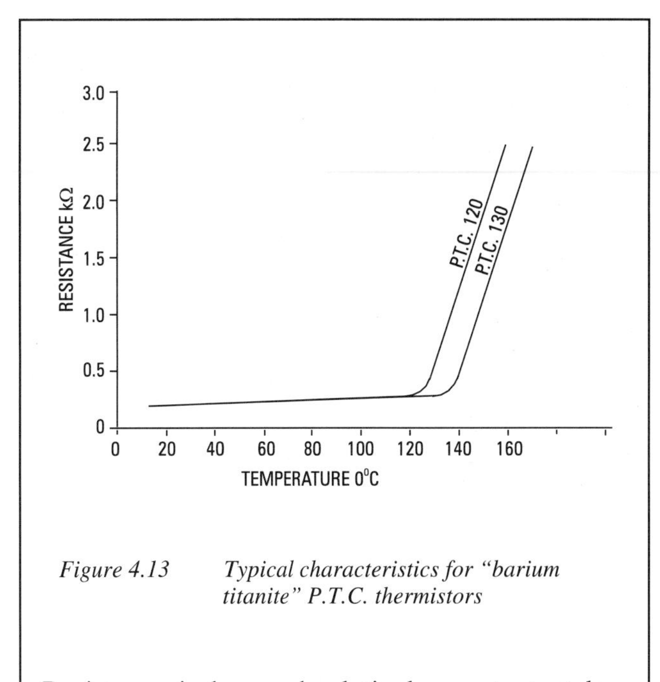

Figure 4.13 *Typical characteristics for "barium titanite" P.T.C. thermistors*

Resistance is low and relatively constant at low temperatures. The rate of increase becomes very rapid at the SWITCHING TEMPERATURE POINT. Above this point the characteristic becomes very steep and attains a high resistance value.

This type of overload detection is initially more expensive for the motor has to be purchased with the thermistors already embedded, but as it is more sensitive and responds faster than other methods, it can be more cost effective in the long run due to the reduction in maintenance costs.

Solid-state motor protection

Recent developments allow the basic theory of current transformers to be used with solid-state electronics, to provide an alternative method of overcurrent protection. The current transformers are used to monitor the current in the cables to the motor and the information is fed back to a relay unit. Circuit diagrams for the current transformers are shown in Figures 4.14, 4.15 and 4.16.

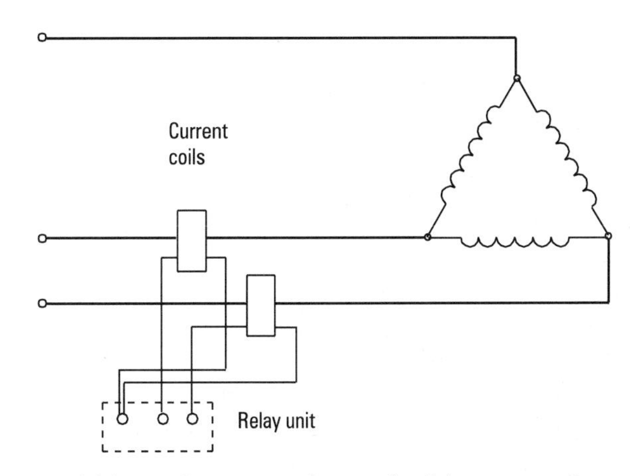

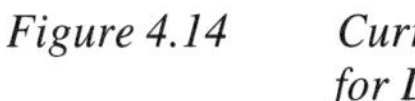

Figure 4.14 *Current transformers for delta connected motor for DOL starting or soft start*

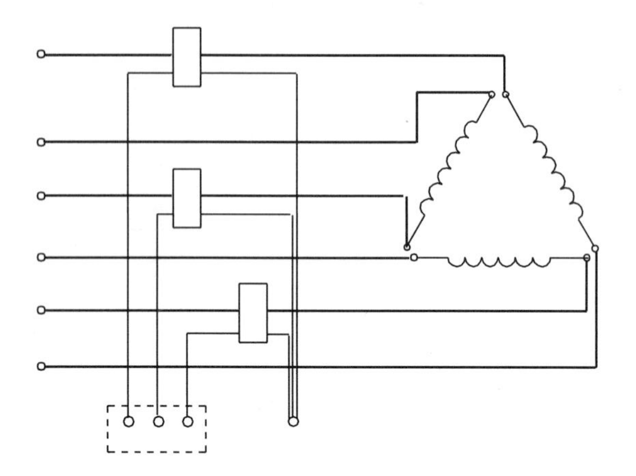

Figure 4.15 *Current transformers in the delta loop of star-delta starter*

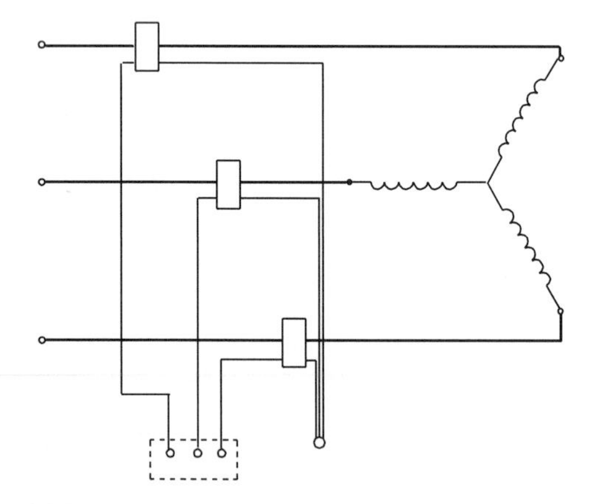

Figure 4.16 *Current transformers for a star-connected starter*

The solid-state relay is designed so that it can be set to meet the requirements of different applications. A current adjustment is provided so that the actual running current of the motor can be set. A trip-delay can be adjusted so that the relay can be set to correspond to the starting duty of the motor. There is also a reset-delay that can be adjusted according to the characteristics of the motor and load to restrict the number of successive starts the motor can make under overload conditions. In the event of a phase-failure or a phase-imbalance the relay can be set to automatically respond.

Methods of no-volt release

When there is a power failure or considerable drop in voltage it is a requirement that the machine will switch off automatically and not switch on again until manually restarted. There are exceptions to this but this is generally the case. Most starters use an electromagnet arrangement to hold the main contacts of the supply switch in place. This is usually connected so that the contacts are held in place when the control circuit is complete. If for any reason, including voltage drop or supply failure, the circuit becomes open, the main contacts disconnect and the motor stops. It can be seen in the following part of the direct-on-line starter that the circuit can not be completed again until the start button is manually pressed.

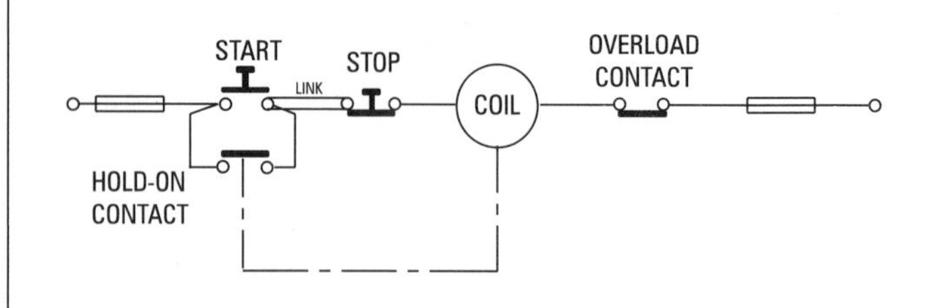

Figure 4.17 *"Hold-in" circuit - when the coil is de-energised the circuit is broken*

Direct-on-line starting

The direct-on-line (DOL) method of starting is by far the most straightforward, but it does have the disadvantages of the high starting currents, as discussed previously. This starter is basically a three-phase contactor that incorporates the necessary safety devices. It is sometimes best to think of the starter being two different circuits, one being the main switching circuit, the other being the control circuit. The main switching circuit is basically a three-phase contactor which is connected into the cables supplying the motor. So that the current can be monitored, devices are connected in series with each of the phases to the motor. An example of this is shown in Figure 4.18.

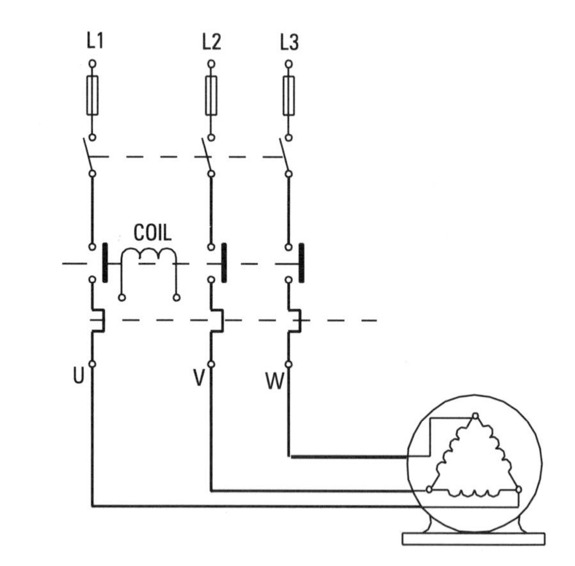

Figure 4.18 *Main circuit contactor with overloads*

The control circuit is used to switch the supply on and off to the operating coil of the contactor. It is made up as shown in Figure 4.19.

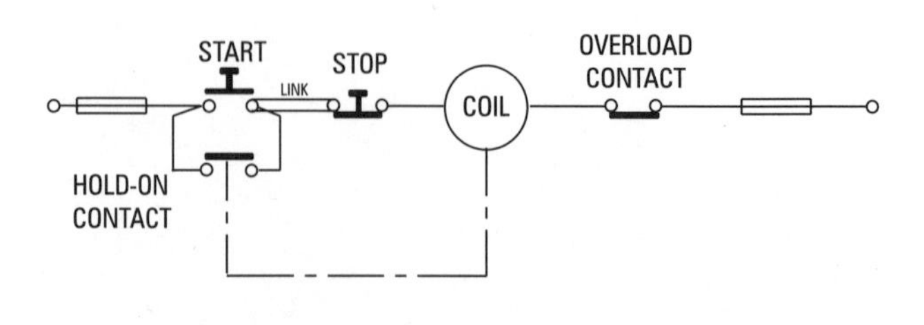

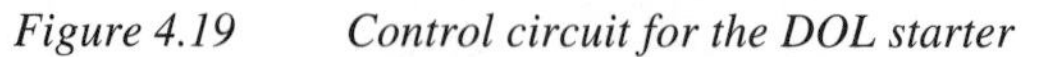

Figure 4.19 *Control circuit for the DOL starter*

When the main circuit and control circuit are put together the complete direct-on-line starter is produced, as shown in Figure 4.20.

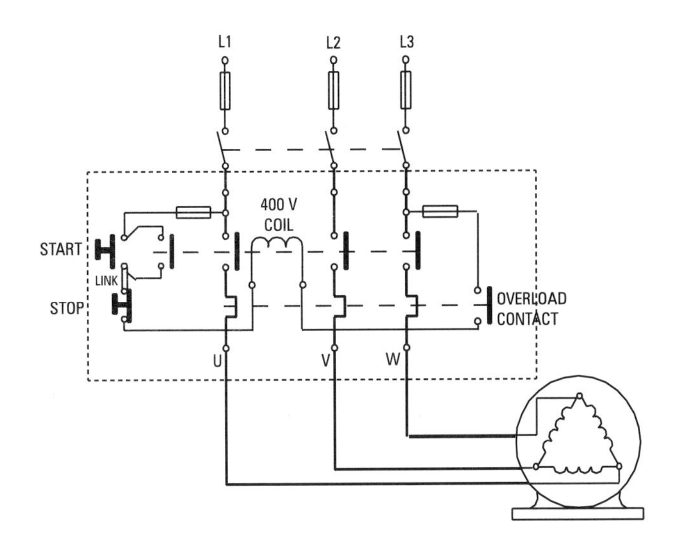

Figure 4.20 *Complete DOL starter with main and control circuits*

Extra stop controls can be connected in series with the operating coil and extra starts can be connected in parallel with the start button. When start and stop controls are required remote from the starter, these can be connected using three conductors and removing the link as shown in Figure 4.21.

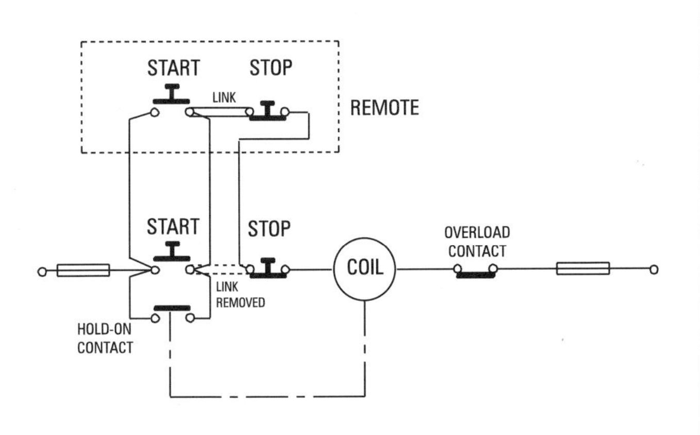

Figure 4.21 *Remote stop start*

Where there are motors driving machinery like long conveyors it may be necessary to have stop buttons at remote places. On some equipment the stop buttons would be replaced by interlocks on safety guards. All of these would be wired in series with the operating coil. Often in these circumstances the start would be a key switch operated by qualified personal. An example of this type of circuit is shown in Figure 4.22.

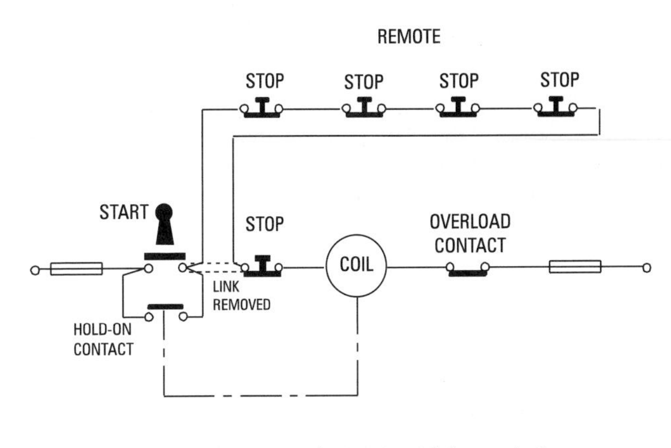

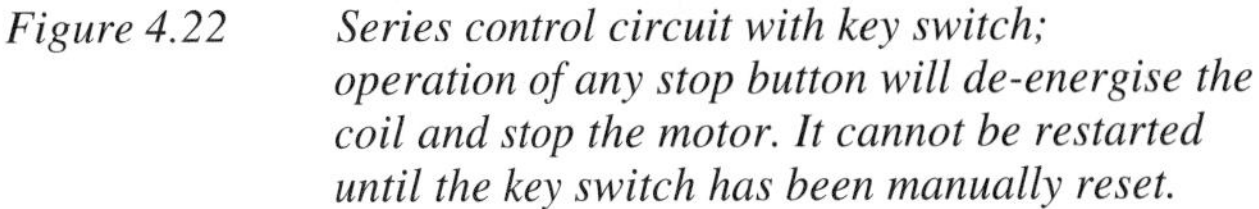

Figure 4.22 *Series control circuit with key switch; operation of any stop button will de-energise the coil and stop the motor. It cannot be restarted until the key switch has been manually reset.*

Where the motor has to be "inched", in other words only switched on when a button is pressed, the control circuit can be adapted using a button with two contacts, one normally open the other normally closed. This is sprung loaded so that it always returns to the same position.

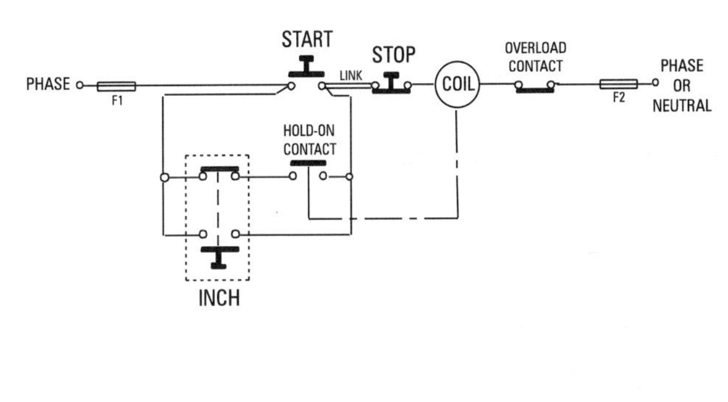

Figure 4.23 *Inch control*

By using these basic control circuits complex switching arrangements can be built up. Where multiprocess equipment is installed it is often necessary to have safety interlocks fitted to removable covers, as well as having elaborate starting and reset arrangements.

Remember

Generally, to stop a machine, normally closed contacts are connected in series with the operating coil. Similarly, to start a machine, normally open contacts are connected in parallel with the hold-in contact.

Reversing the direction of rotation

It was shown earlier that by changing over any two phases of a three-phase supply the direction of rotation of a motor can be reversed. This can be carried out manually by actually disconnecting and changing over the connections inside the motor terminal box. On occasions, though, it is necessary to be able to reverse a motor at the push of a button. This can be achieved by using a combination of direct-on-line starters (Figure 4.24).

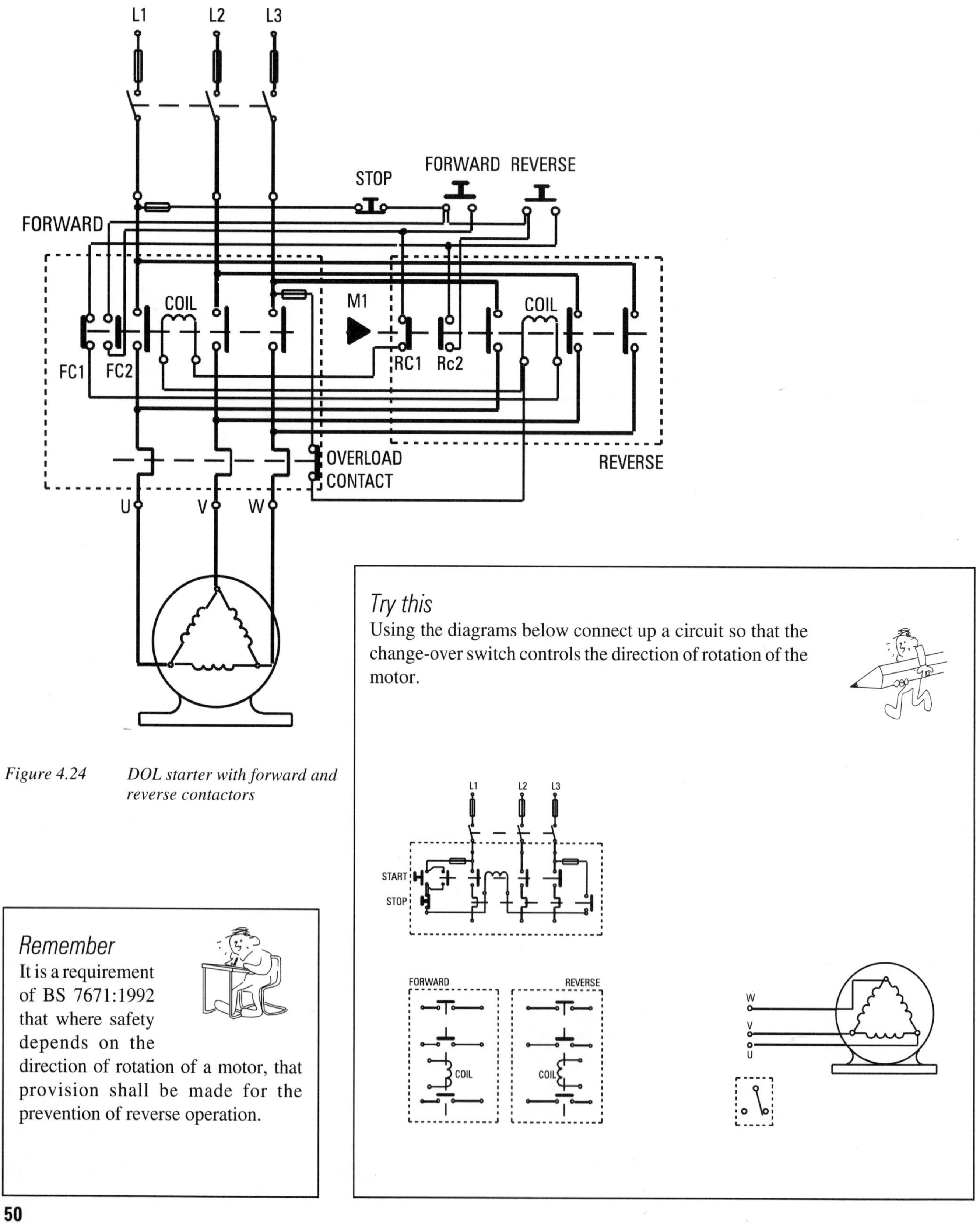

Figure 4.24 *DOL starter with forward and reverse contactors*

Try this

Using the diagrams below connect up a circuit so that the change-over switch controls the direction of rotation of the motor.

Remember

It is a requirement of BS 7671:1992 that where safety depends on the direction of rotation of a motor, that provision shall be made for the prevention of reverse operation.

Reduced current starting

Where the rating of a motor exceeds the permitted maximum, methods of starting that reduce the initial current must be employed.

Star-delta starting

If a motor is started with the windings connected in star, the current is reduced to approximately 33% of the direct-on-line value. To be able to do this, the motor must have all six winding connections available. These must be clearly identified so that the start and end of each winding can be identified.

So that the motor can be started with the windings connected first in star and then switched to delta a special starter has to be used. This may be a manually operated switch or automatic.

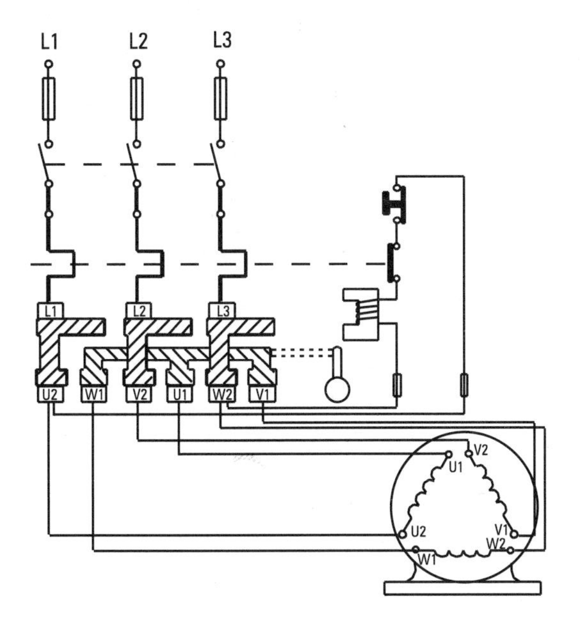

Figure 4.25 *Started in star*

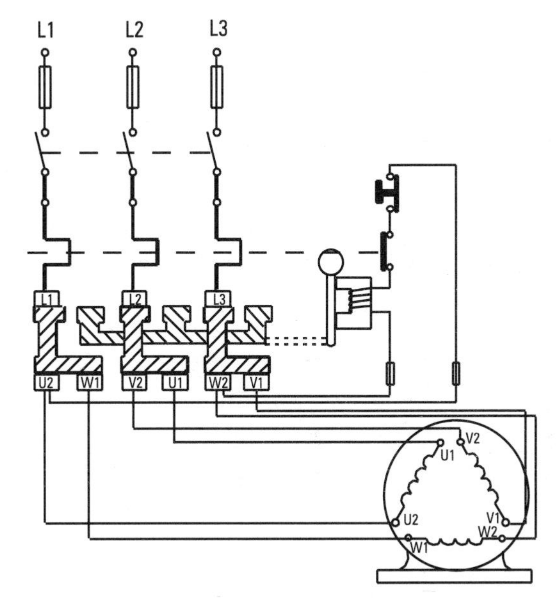

Figure 4.26 *Run in delta*

Manually operated switch

Operation

Move the change-over switch downwards to the "star" starting position. (Supply lines L1, L2 and L3 are connected to U2, V2 and W2 respectively, and U2, V1 and W1 are connected to star.)

Allow sufficient time for motor to run up to speed.

Move the change-over switch upwards to the "delta" running position (L1 is connected to U2 and W1, L2 is connected to V2 and U1, L3 is connected to W2 and V1) and leave the switch in this position.

The no-volt coil energises and holds the change-over switch in the "delta" position by an electromagnetic mechanism. The motor continues running.

Press the stop button, the no-volt coil de-energises, the switch mechanism releases, the switch contacts open and the motor stops.

The no-volt coil de-energises also when:

- an overload situation occurs
- there is a supply failure
- the supply voltage falls below a certain value (undervoltage situation)

Consequently the motor stops and cannot restart until the change-over switch is moved downwards again to the star starting position by the operator.

Automatically operated switch

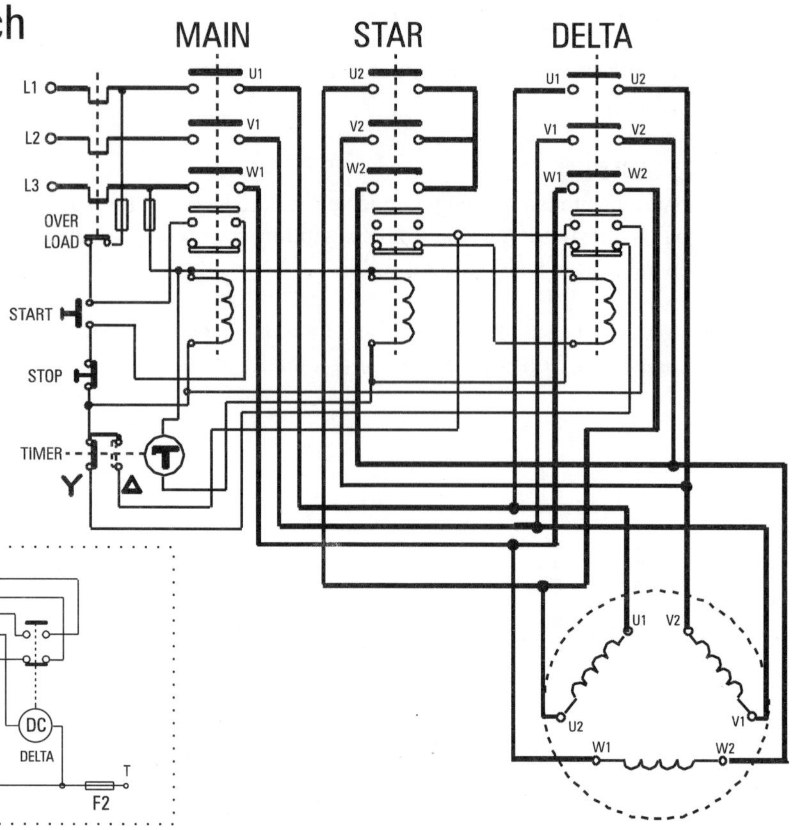

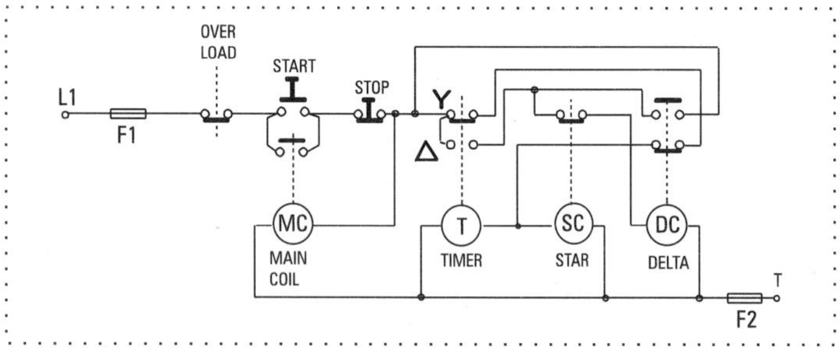

Figure 4.27 Automatic star-delta starter

Operation

Press the start button and the main contactor coil energises.

The star contactor coil and timer (**T**) both energise via the timer contact (**Y**) and "delta" interlock contact.

The main contacts close and connect the three-phase supply lines to the motor winding terminals U1, V1 and W1.

The star contacts close and connect motor winding terminals, U2, V2 and W2 together to form the "star point".

The motor starts to run on reduced voltage with 230 V across each winding.

Release the start button and the main contactor coil remains energised via the hold-on contact and the timer contact (**Y**). The motor continues to run on reduced voltage.

The timer "times-out" and switches the changeover contact from the **Y** position to the Δ position.

The star contactor coil de-energises and the delta contactor coil energises via the timer contact Δ and the star interlock contact.

The delta contacts close and connect the following motor winding terminals together U1 to V2, V1 to W2 and W1 to U2 to form the delta connection of the windings.

The motor continues to run on full supply voltage with 400 V across each winding.

The timer de-energises (resets) to the star position, and the delta contactor coil remains energised via its own hold-on contact.

Press the stop button and the main and delta contactor coils de-energise, the main contacts open and the motor stops.

Auto-transformer starting

Where motors only have delta connections available, with three terminals provided, it is not possible to use star-delta methods of starting without access to inside the motor. In these cases, and where star delta starting does not give sufficient starting torque, an auto-transformer can be used. There are usually several taps taken from the transformer. An example would be 50, 60 and 75% of the line voltage. At 50% the current is only

$$\left(\frac{1}{2}\right)^2, \text{ or } 0.25,$$

of the full load current, and as the torque produced at values below this is very low, the 50% tap is usually the lowest. After the motor has accelerated using one of the tapped voltages, the motor windings are connected direct across the main supply voltage.

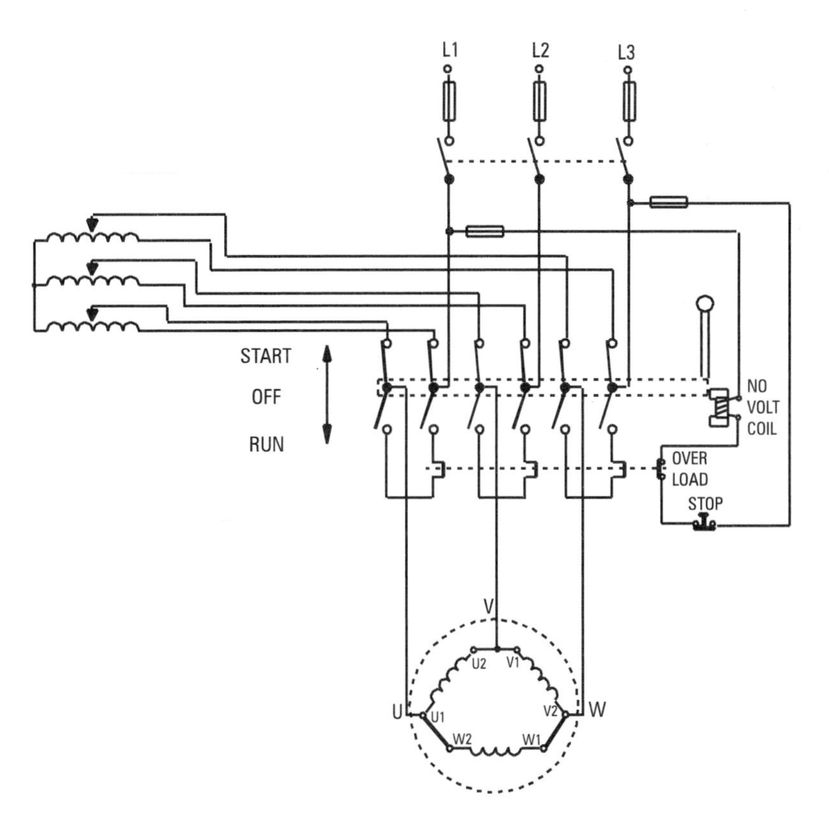

Figure 4.28 *Auto-transformer starter shown in the start position with the transformer tappings at 50%.*

Wound rotor (slip-ring) induction motor

It was seen earlier in this chapter that if the resistance of the rotor is increased the starting torque can be increased and the starting current reduced. The wound rotor slip-ring motor takes advantage of this by using external rotor resistances during the starting process. When the motor has reached full speed the resistances are shorted out so that the motor can run at minimum slip and increased efficiency. The wound rotor has three sets of windings connected in star and brought out to slip-rings for connection to the external resistances.

Note: Starting torque is usually kept down to about 150% full-load-torque, with a starting current of about 150% of full-load-current. This is done to reduce the disturbance to the mains supply and the "shock" to the driven machine.

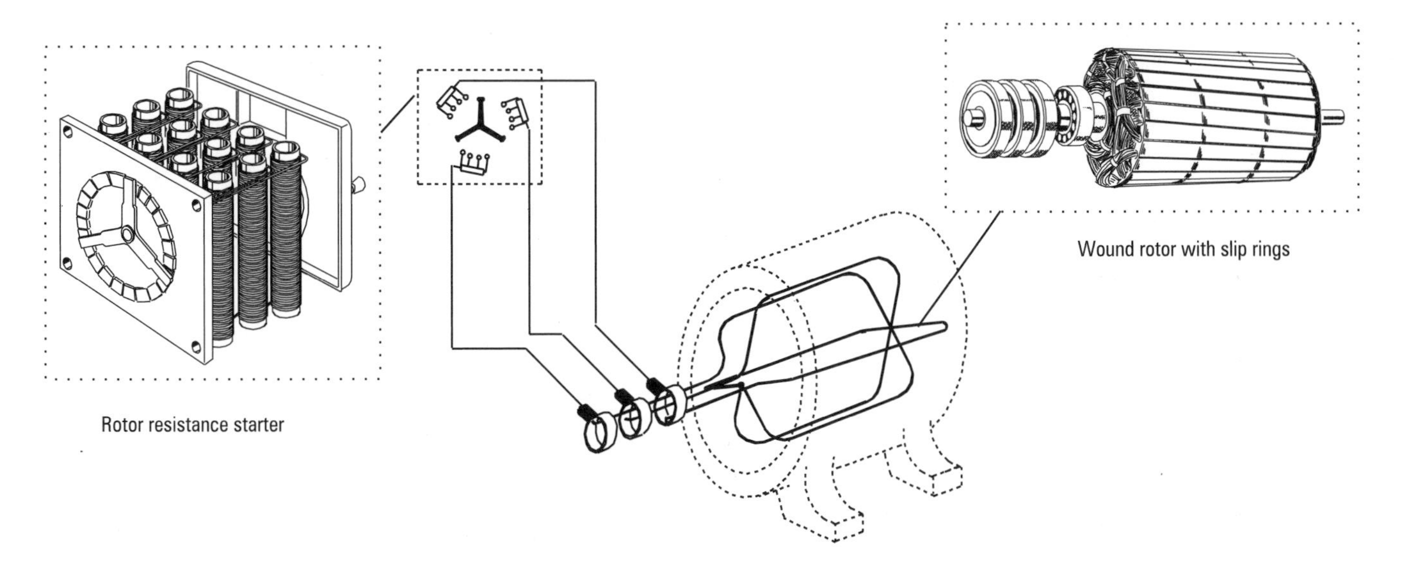

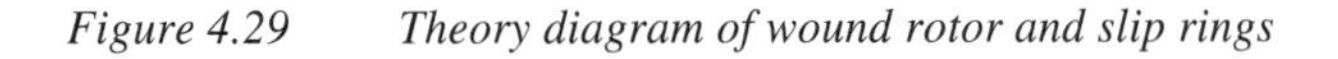

Figure 4.29 *Theory diagram of wound rotor and slip rings*

Solid-state "soft start" controller/starters

These devices provide an alternative to the starting methods already shown. They are designed so that they not only limit the inrush current when the motor is first switched on, but also control smooth acceleration and torque build up to meet the requirements of the load. There are several variations on types of solid state starters but all have similarities.

In most types a control board is used to supply firing pulses to a thyristor stack in a regulated sequence. The time may be controlled by the user or automatically depending on the type. This provides a load voltage to the motor to suit the required application. Figure 4.30 shows the parameters of the starting profile.

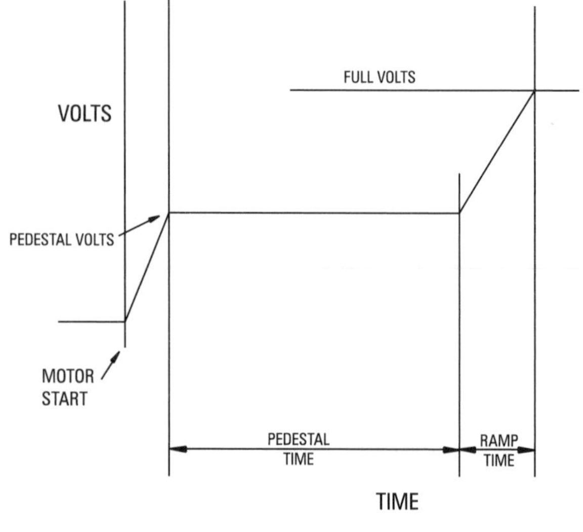

Figure 4.30

When the motor is switched on the voltage to it is increased to a predetermined level which is known as the pedestal voltage. It is maintained at this level for what is known as the pedestal period of time, so that the starting current is controlled whilst the voltage is sufficient to allow the motor to generate the necessary breakaway torque. After this predetermined pedestal time the controller automatically moves on to the ramp stage. The voltage now increases allowing the motor to accelerate smoothly up to full speed in the running condition with the full voltage supplied to it. Typical current/time characteristics for this starting cycle are shown in Figure 4.31 and it can be seen that the starting current is kept to about 200% of full load current, unlike the DOL starter which can be up to 600%.

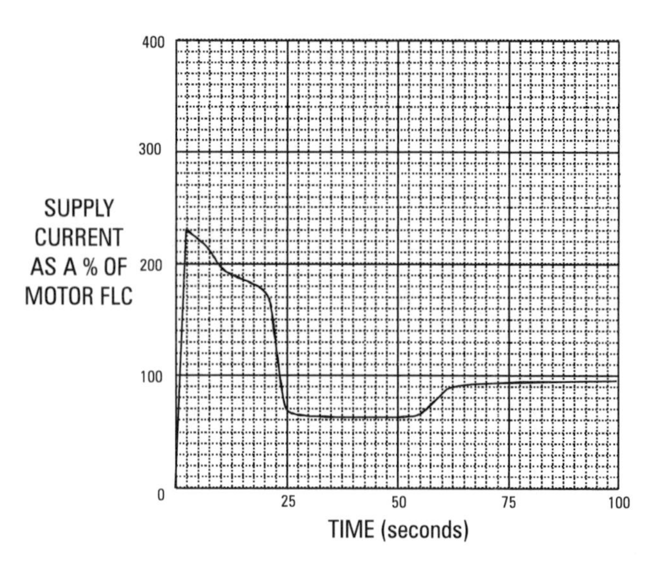

Figure 4.31 *Current/time characteristics*

When the motor is switched off the motor can shut down instantaneously or alternatively the control board may be selected. In the latter case the controller will automatically ramp down the voltage to give a soft stop to the motor. Figures 4.32 and 4.33 show the main circuit diagram and control circuit diagram, for this arrangement.

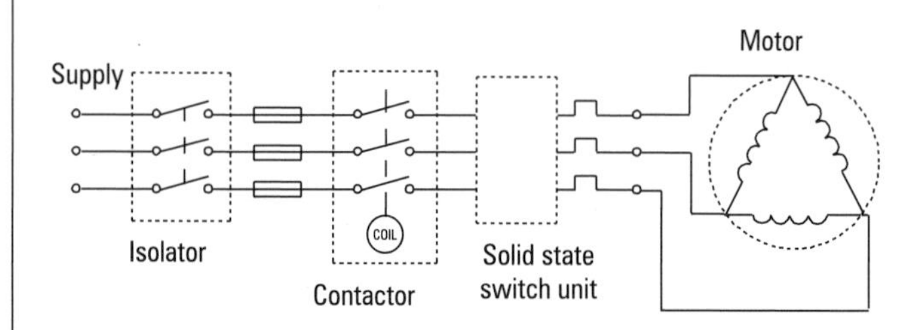

Figure 4.32

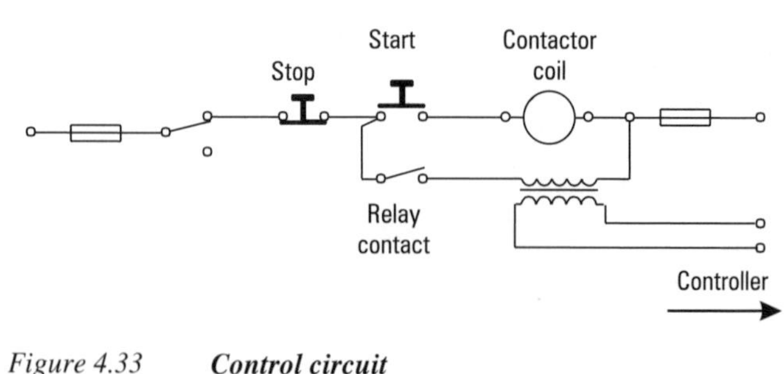

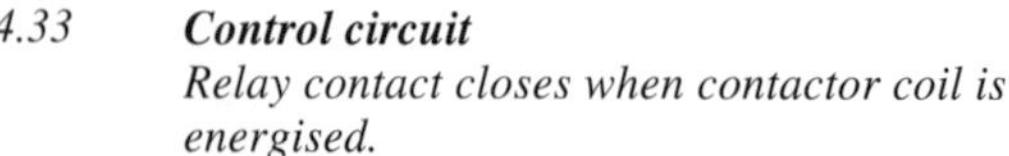

Figure 4.33 ***Control circuit***
Relay contact closes when contactor coil is energised.

Single-phase motors

When looking at how a three-phase motor works it was seen how necessary it was to have a rotating magnetic field. With three-phases this is available automatically, but on single phase this has to be created to get the motor started. Once the rotor is spinning, it will continue to do so even though the stator is only supplied with a single phase. This is due to the fact that the emf in the stator induces an emf in the rotor through mutual inductance. This emf in the rotor creates its own pulsating magnetic field which is 90° behind that in the stator, both in time and space. There are several different methods used to create the phase displacement required to produce rotation. Each method has its own characteristics and applications.

Split-phase motors

The stator consists of two sets of windings, one the main winding, the other an auxiliary winding which is there for starting purposes only. The auxiliary winding is made up of fine wire which has a high resistance, and is embedded less deeply into the laminated iron stator core than the main winding, which gives it a comparatively low reactance. As the main winding has a low resistance and high reactance there is a phase displacement between the two. This is enough to create a moving magnetic field to get the rotor turning. Figures 4.34 and 4.35 show the connection diagram and resulting phasor diagram for this arrangement.

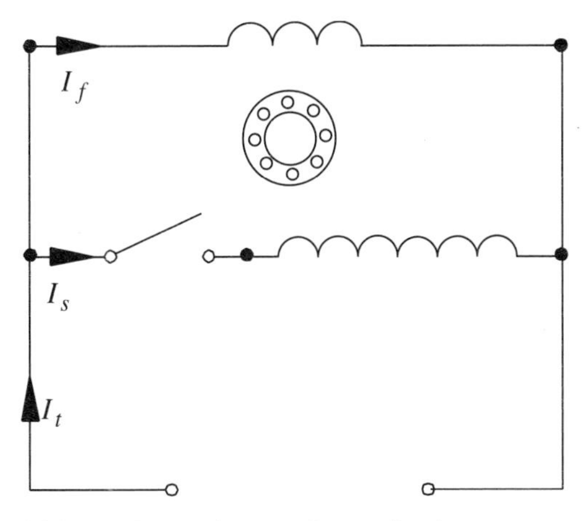

Figure 4.34 *Circuit diagram for a split-phase motor*

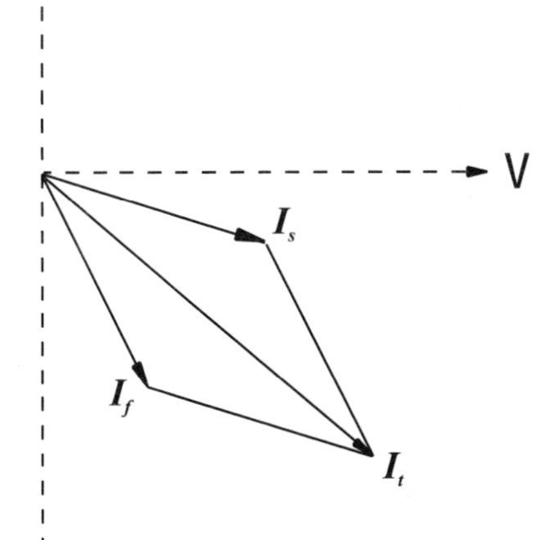

Figure 4.35 *Phasor diagram for the split-phase motor*

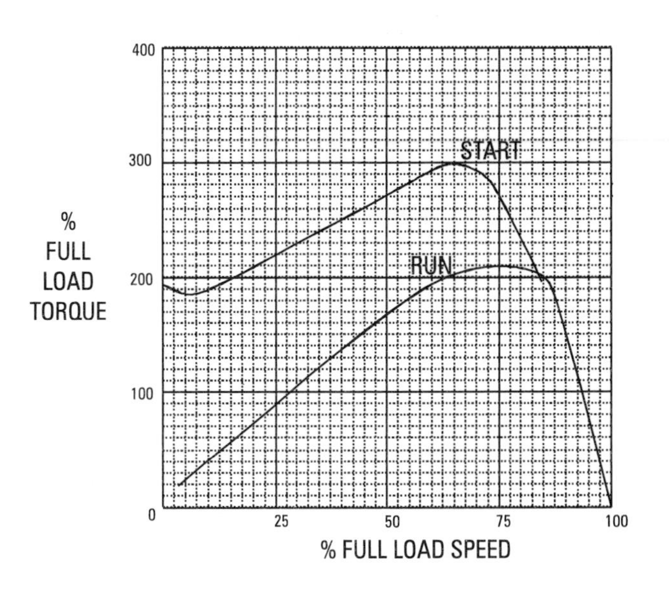

Figure 4.36 *Torque/speed characteristics of a split-phase motor*

The auxiliary winding is working on a high current density due to the low resistance that is required. This means that it must be switched out of the circuit as soon as the rotor reaches about 75% of full speed. This switch is usually incorporated in the motor and works on the centrifugal force created by the rotation of the motor shaft.

This type of motor tends to be used for light loads up to 370 W which are not continually stopped and started. The comparatively small phase angle between the two windings mean that the motor is not suitable for starting on load.

Capacitor start – induction run motors

To create a greater phase angle between the two windings used in a split-phase motor, a capacitor is connected in series with the auxiliary winding. This has the effect of reducing the inductive reactance of the auxiliary winding to a low value and making its current lead that of the main winding by almost 90°. The capacitor used for this is usually a short time rated, high value electrolytic type. Figures 4.37 and 4.38 show the connection and phasor diagrams for this arrangement.

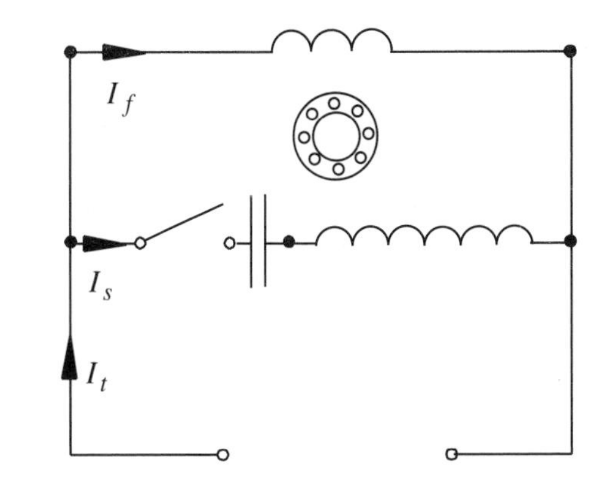

Figure 4.37 *Circuit diagram of capacitor start motor*

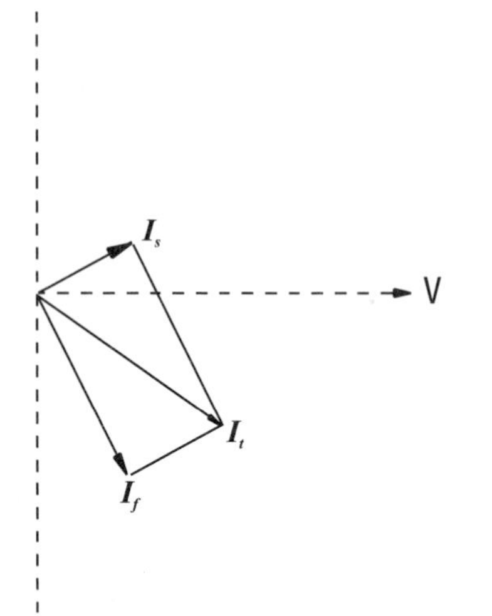

Figure 4.38 *Phasor diagram for the capacitor start motor*

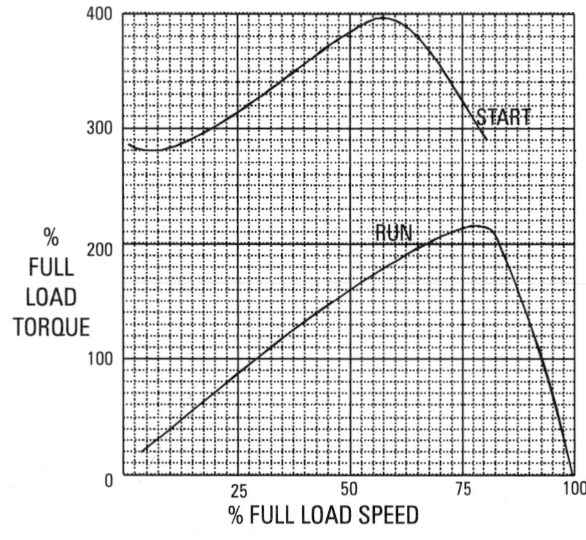

Figure 4.39 *Torque/speed characteristics for a capacitor start motor*

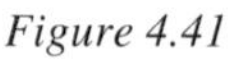

When the motor has reached about 75% of full speed a centrifugal switch again cuts out the auxiliary winding and hence the capacitor.

The greater phase angle means that this type of motor can be started on load and is often used on compressors for refrigeration.

Capacitor start and run motors

There are many similarities between this type of motor and the capacitor start type. There are also some important differences. Although the capacitor is connected in series with the auxiliary winding there is no centrifugal switch to disconnect this circuit and it remains active all the time. As a result of this, the capacitor must be a low value continually rated type. The continuous large phase shift between the windings give this type of motor a running performance similar to that of a three-phase machine. The starting however, is something less than that of a capacitor start motor. Figures 4.41 and 4.42 show the relevant diagrams.

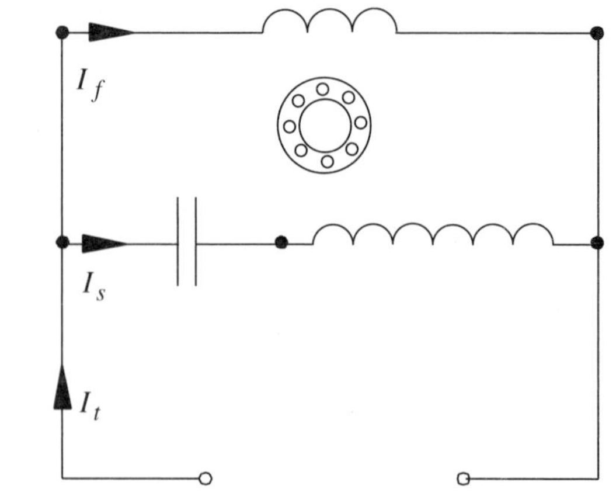

Figure 4.40 *Circuit diagram of capacitor start and run motor*

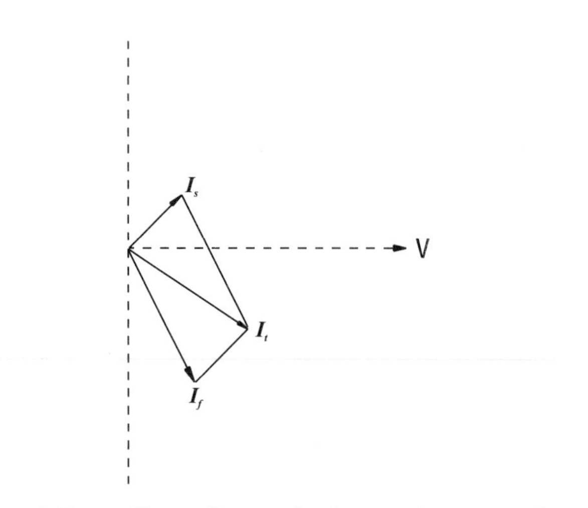

Figure 4.41 *Phasor diagram for the capacitor start and run motor*

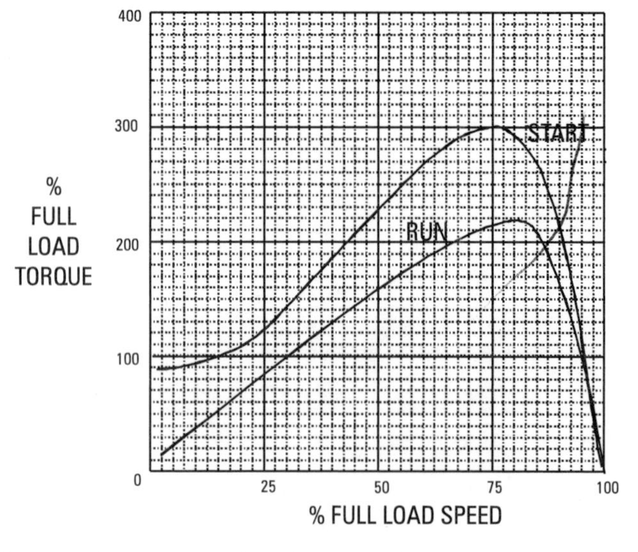

Figure 4.42 *Torque/speed characteristics for a capacitor start and run motor*

These motors are often used for fans and central heating circulating pumps.

Capacitor start – capacitor run motors

This type of motor is a combination of the previous two types, using two capacitors connected into the auxiliary winding. One capacitor is a high value short time rated type and the other a continuously rated low value type. When starting the motor has both capacitors connected in parallel with each other and in series with the auxiliary winding. This gives a large capacitance resulting in an increased phase angle. When the rotor reaches about 75% of full speed the centrifugal switch disconnects the high value short time rated capacitor, leaving the other in series with the auxiliary winding to continue working. This gives a motor with a starting performance similar to the capacitor start motor but with the running performance of the capacitor start and run type. Figures 4.43 and 4.44 show the relevant diagrams.

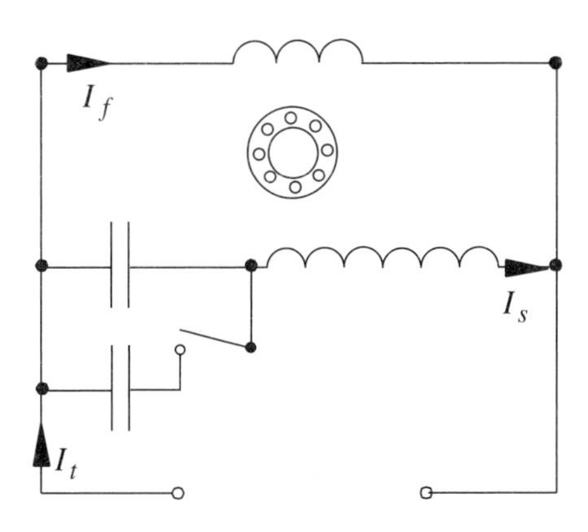

Figure 4.43 *Capacitor start - capacitor run motor and circuit diagram*

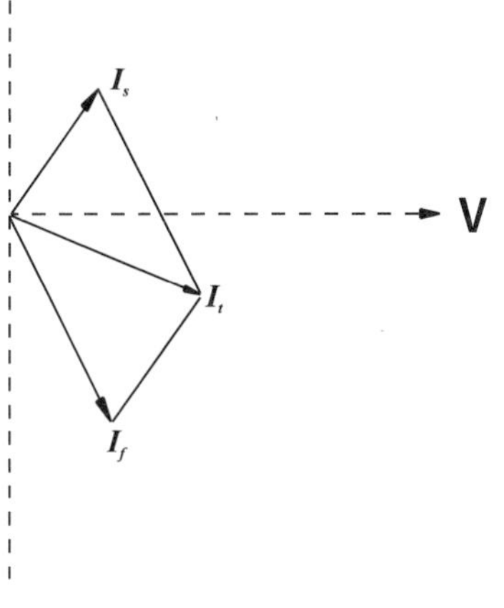

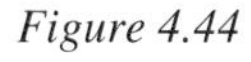
Figure 4.44 *Phasor diagram for the capacitor start - capacitor run motor*

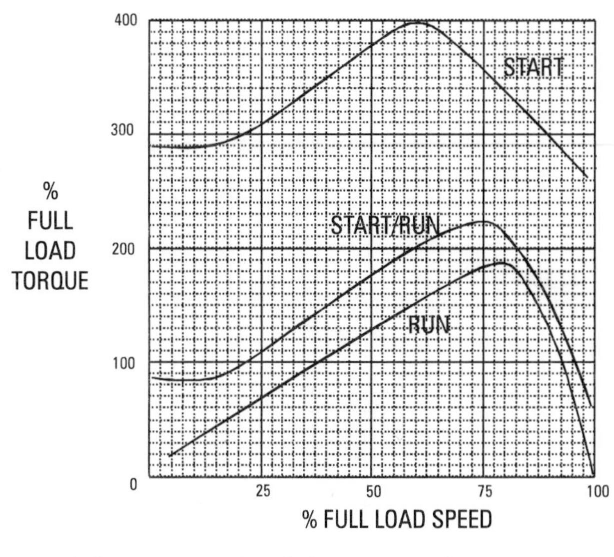

Figure 4.45 *Torque/speed characteristics for a capacitor start - capacitor run motor*

Shaded pole motor

The shaded pole motor is usually easily identifiable from its construction. It is an induction motor with a cage rotor, as shown in Figure 4.46.

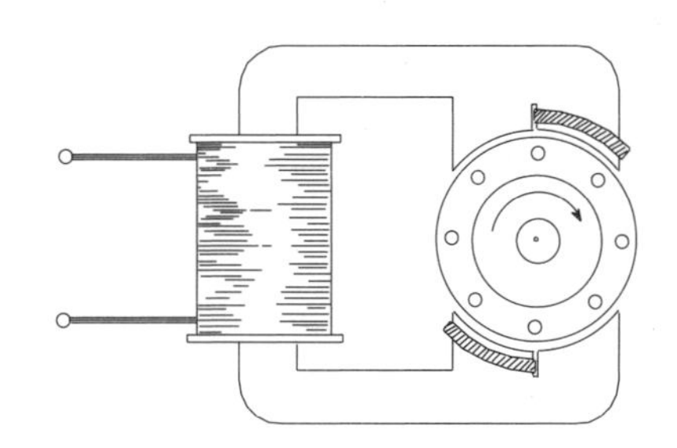

Figure 4.46 *Shaded pole motor*

The stator has two salient poles each with a slot cut into the faces. A short circuited copper ring is fitted around a portion of each pole. When a single phase supply is connected to the stator coil the pulsating magnetic flux causes a transformer action which induces currents in the shading rings. These currents cause the flux in the shaded portion of the poles to lag that in the rest of the poles. This sets up a rotating field which causes the rotor to turn. The starting torque for this type of motor is very low and the continuous losses in the shading rings mean the efficiency is poor. Figure 4.47 shows typical speed/torque characteristics for this type of motor.

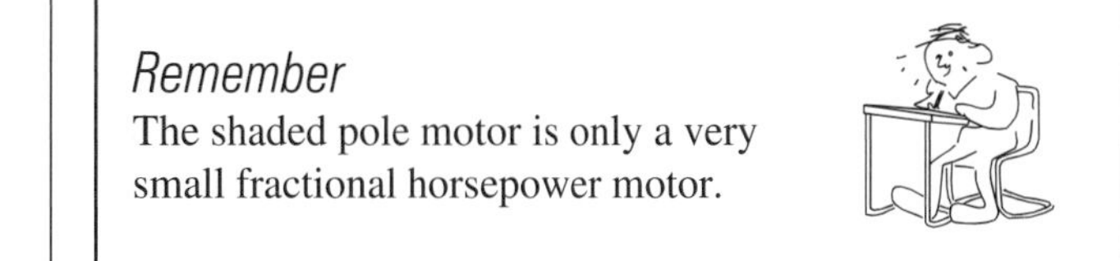

Remember

The shaded pole motor is only a very small fractional horsepower motor.

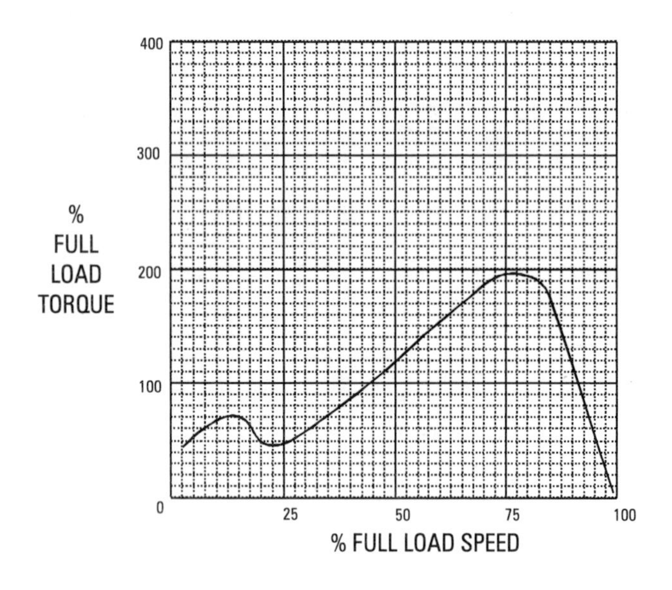

Figure 4.47 *Torque/speed characteristics of a shaded pole motor*

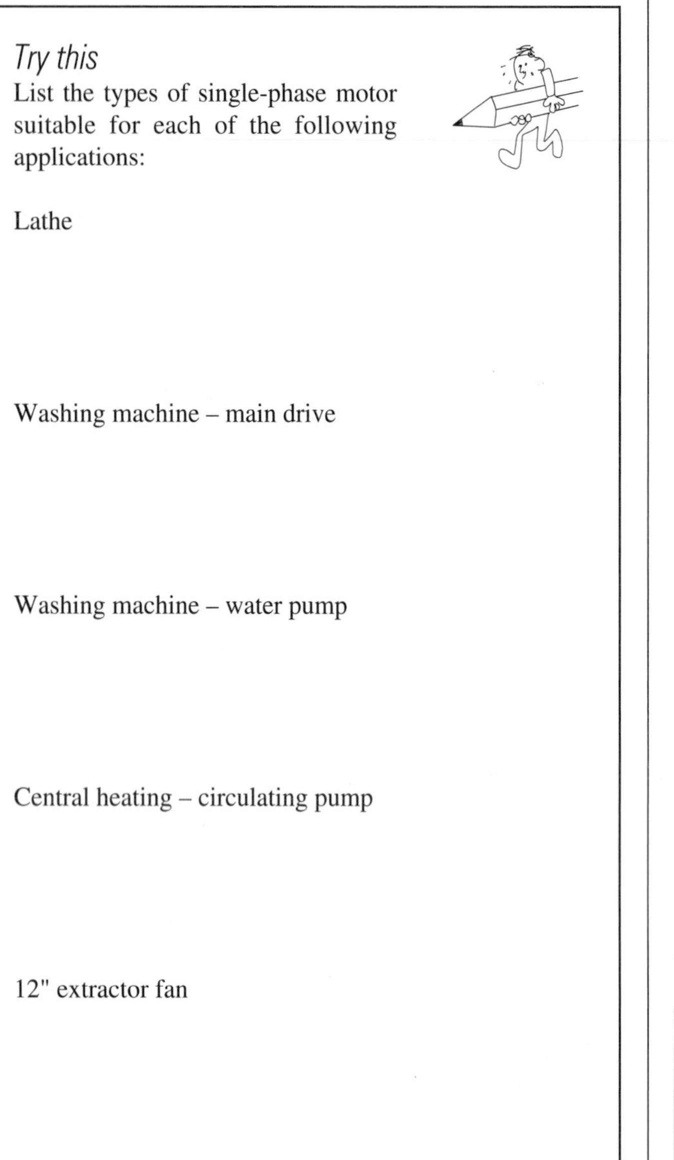

Try this

List the types of single-phase motor suitable for each of the following applications:

Lathe

Washing machine – main drive

Washing machine – water pump

Central heating – circulating pump

12" extractor fan

Universal motors

This type of motor is basically the d.c. series motor, but with a laminated yoke to prevent overheating due to eddy currents.

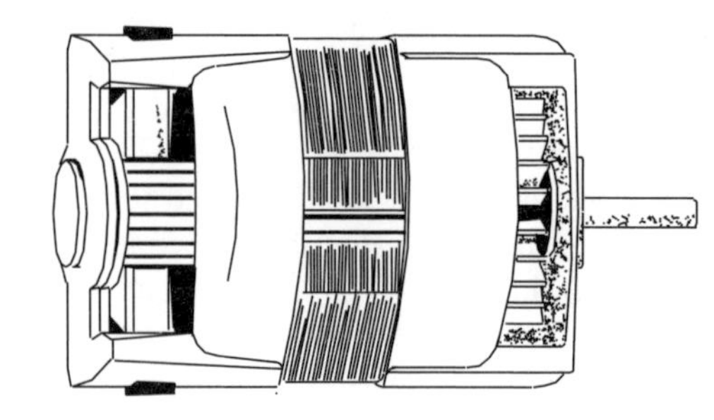

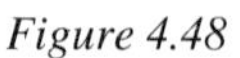

Figure 4.48

When it is used on a.c. the continual switching action of the brushes on the armature can cause radio interference. This means that a suppression unit is often found connected as shown in Figure 4.49.

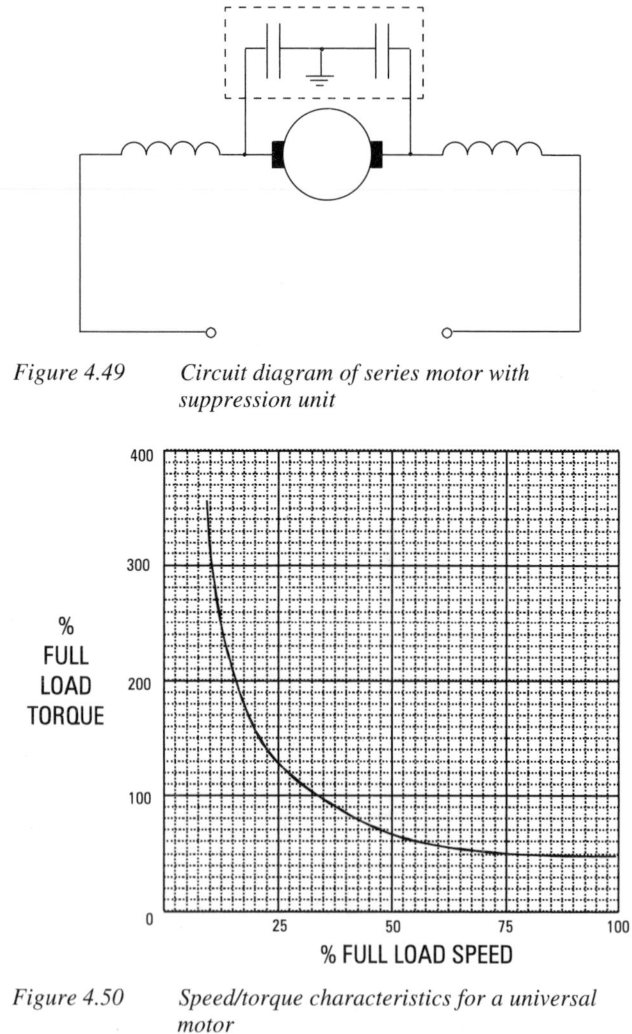

Figure 4.49 *Circuit diagram of series motor with suppression unit*

Figure 4.50 *Speed/torque characteristics for a universal motor*

Remember

The universal motor can run at very high speeds (typical 8000–12000 r.p.m. maximum)

Speed control

It was shown earlier that the speed of an a.c. motor is determined by

$$n = \frac{f}{p}$$

where

n = speed in revs/sec

f = frequency (Hz)

p = number of pairs of poles

Altering either the number of poles or the frequency, effects the shaft speed of the motor.

Number of poles

Where a motor is to be used on two fixed speeds it can be constructed with two sets of windings. One set would be wound with the appropriate number of poles for one speed and the second windings would be completely separate and have the number of poles for the other required speed. An example of where this has been used on single-phase machines is the domestic washing machine. The main drive motor is run as a four-pole motor, at about 1450 rev/min, for washing and then switched to a two-pole winding and run at approximately 2900 rev/min for the spinning cycle. The physical size of motors constructed in this way are larger than single speed motors. They are however, smaller and more efficient than motor gearbox arrangements.

There are many other examples where motors are constructed so that more than one set of poles can be used to vary the shaft speed.

Frequency

As already seen, a.c. motors naturally run at speeds 1% to 6% below the synchronous speed. As synchronous speed is determined by the supply frequency it follows that varying the supply frequency automatically changes the rotor speed of the motor. The methods of creating a change in the supply frequency to motors has changed considerably over the years. Until the development of power electronics, rotary converters were used. Rotary frequency changers have tended to be 2-pole motors driving multipole alternators. These have disadvantages of being moving machines that make noise, take up large spaces and require regular maintenance. In recent years these have been superseded by electronic methods. The development of semi-conductor devices for use in power circuits has been put together with micro electronics to produce efficient frequency controllers.

The basic theory of electronic frequency control is that an a.c. supply is taken and converted into d.c. As the a.c. supply to the unit is constant, the resulting d.c. is also constant. This constant d.c. is supplied into an inverter bridge which switches it so that an artificial a.c. is produced. The frequency of this "a.c." can be adjusted over a wide range to give speed control.

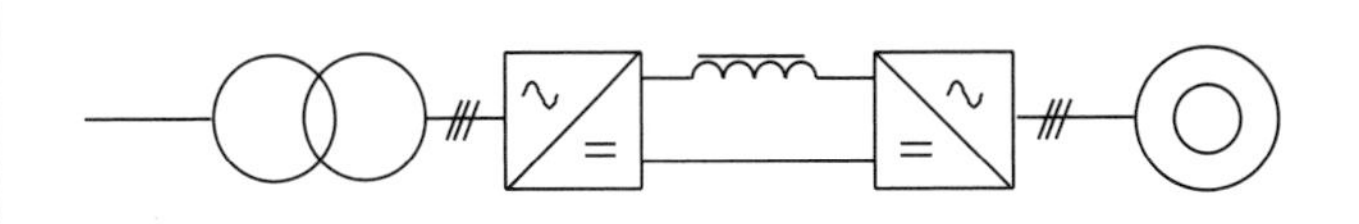

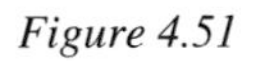

Figure 4.51

There are several different methods that may be used for the switching operation to produce the artificial a.c. Three of these are

- pulse amplitude modulation (PAM)
- pulse width modulation (PWM)
- current source inverter (CSI)

The pulse amplitude modulation system uses pulses of voltage to the motor circuit to create a current that is crudely sinusoidal in shape. The quality of the shape of the current wave form is related to the number of pulses of voltage each cycle. An example of the voltage pulses and resulting current wave form is shown in Figures 4.52 and 4.53.

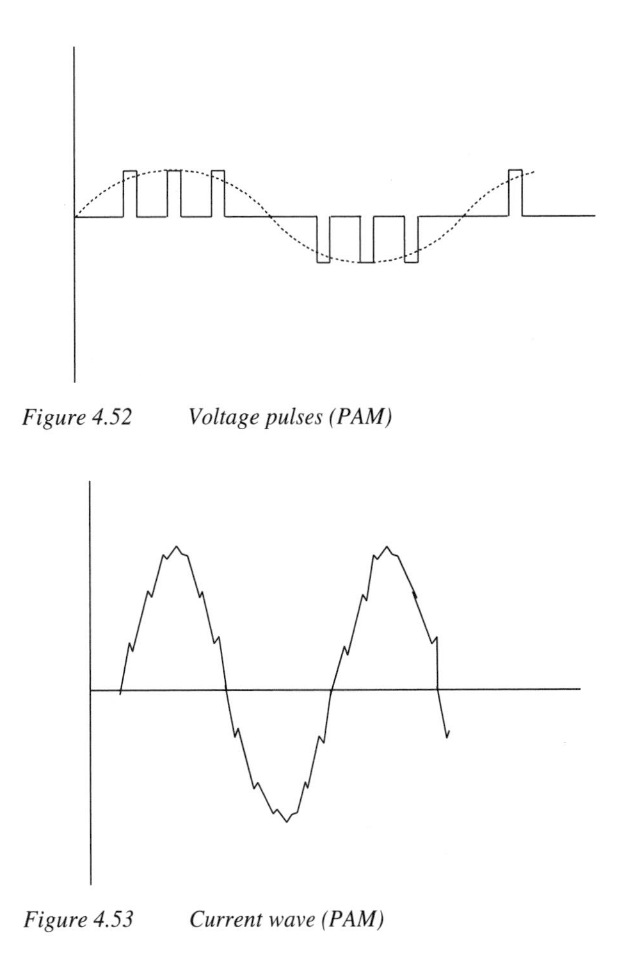

Figure 4.52 *Voltage pulses (PAM)*

Figure 4.53 *Current wave (PAM)*

When only a few pulses are used the non-sinusoidal wave form that results, generates harmonic frequencies which create a "hunting" effect on the motor and also increased heat losses.

The pulse width modulation system also uses pulses of voltage to control the motor current. With this system the d.c. voltage is switched so that the width of the pulse is related to the shape of the sine wave it is trying to emulate. Figure 4.54 shows a typical voltage pulse arrangement.

The resulting current wave is not unlike that of the PAM system for a similar number of pulses. Like the PAM system the more voltage pulses there are, the better the resulting current wave form and less the losses. There are ways of controlling PWM systems so that the harmonics are automatically cancelled and the losses are still further reduced.

The current source inverter is different from the other two systems as it uses a constant current and variable voltage. The switching of the inverter can be designed to create either constant width or variable pulse characteristics.

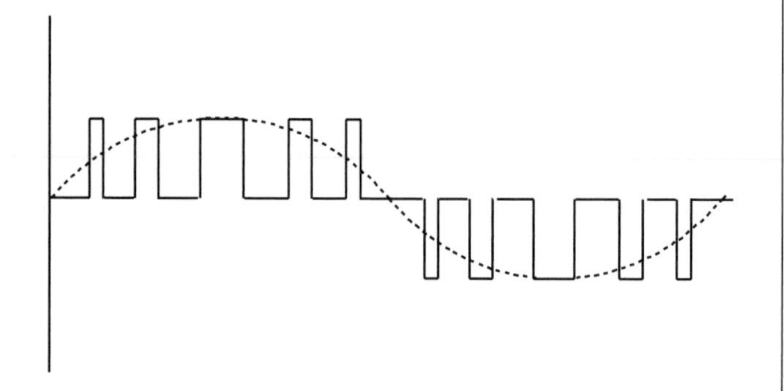

Figure 4.54 Voltage pulses (PWM)

When installing inverters it is important to keep the cables to the motor as short as possible and separate from other power cables. Consideration may also have to be given to earth fault leakage protection as it is not recommended that residual current devices are used.

Wound rotor motors

When looking at speed control using wound rotor motors there are two completely different systems to be considered. First there is the three-phase motor system which uses the Schrage motor and secondly single-phase motors which uses the universal series motor.

Three-phase

In the past where a high degree of speed control was required on special equipment, a motor known as the Schrage motor, was used. The development of electronic methods of speed control have, to a large extent, replaced the need for this complex wound machine. Figure 4.55 shows a typical circuit diagram of the internal connections of a Schrage type motor.

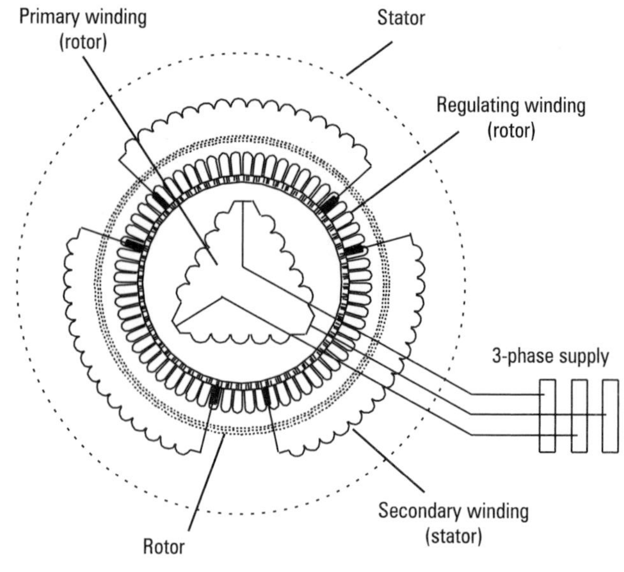

Figure 4.55 Rotor-fed a.c. commutator motor (Schrage motor)

This type of motor consists of three sets of windings, two on the rotor and one on the stator. The rotating magnetic field is created in three-phase delta connected windings on the rotor. The supply to these is via three-slip rings mounted on the drive shaft. The second windings on the rotor are the regulator windings which are taken to segments of a commutator similar to those of a d.c. machine. The stator consists of three sets of windings with each end connected to brushes which are in contact with the commutator. The brushes are arranged so that one from each winding is fixed whereas the others can all be adjusted together around the commutator.

When the motor is first switched on, and before it rotates, the rotating magnetic field sweeps past the stator windings and the brushes, at synchronous speed. As the rotor begins to gain speed, the frequency in the stator windings and the brushes, is relatively reduced. When the motor runs at synchronous speed there is no relative movement between the rotating magnetic field and the stator windings, including the brushes. This can be achieved by adjusting the brushes so that the ends of each coil are on the same commutator segment and therefore shorted out. If now the movable brushes are adjusted an e.m.f. can be taken from the d.c. winding and supplied to the stator windings. When the brushes are adjusted in one direction the e.m.f. supplied will create a magnetic field that assists that of the rotating field. This will have the effect of increasing the rotor speed and hence the shaft speed. Adjustment of the brushes in the other direction will create an opposing force which will slow the rotor down and reduce the shaft speed.

The position of the brushes can be adjusted manually by turning a handwheel mounted on the side of the machine, or they can be adjusted remotely by means of a small servo motor.

Pole change motor

The basic idea of a pole change motor is to provide a different number of poles on the same stator to change the speed of the motor.

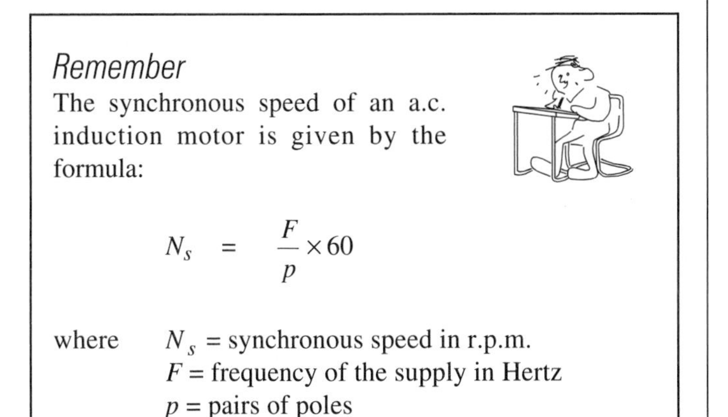

Remember

The synchronous speed of an a.c. induction motor is given by the formula:

$$N_s = \frac{F}{p} \times 60$$

where N_s = synchronous speed in r.p.m.
F = frequency of the supply in Hertz
p = pairs of poles

One method of obtaining two speeds from a cage-rotor type a.c. induction motor is to have two separate stator windings with a different number of poles. The change in speed is normally achieved by switching between the two sets of windings using contactors.

An alternative method is to provide a different number of poles off a single stator winding.

Example

If each phase of a three-phase stator winding is made up of four coils arranged around the stator core, then by switching between series and parallel connection of the coils a different number of poles can be achieved.

Figure 4.56 shows that when the coils are connected in **series**, **eight poles** are achieved, thus providing **low speed operation**, whilst the **parallel** connection of the coils creates **four poles** for **high speed operation**.

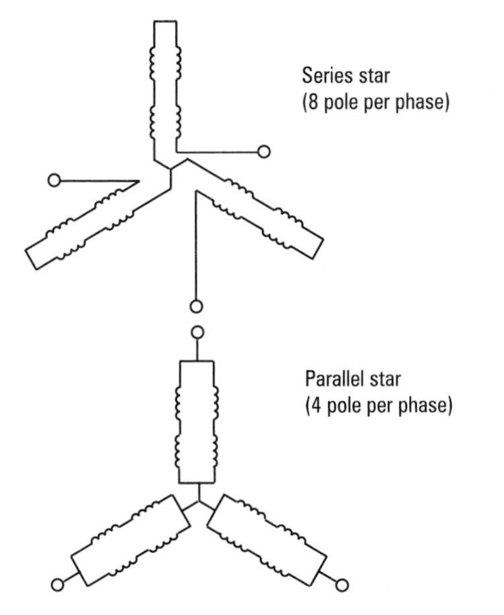

Figure 4.56a *3-phase connection diagram for 8/4 pole changing*

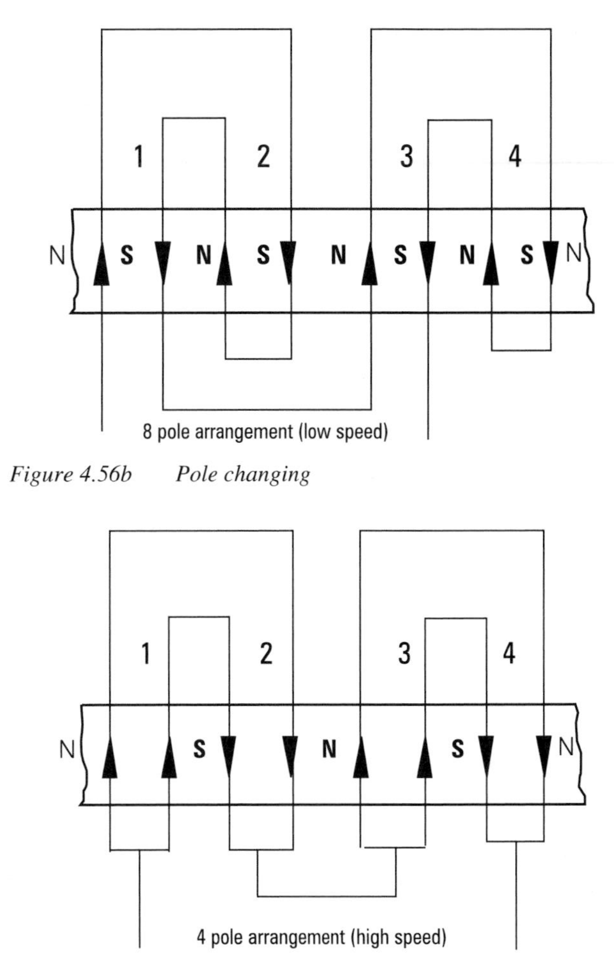

Figure 4.56b *Pole changing*

Figure 4.56c *Pole changing*

Applications

This type of motor does not provide smooth speed variation, but they are widely used for machine tools, lifts and cranes (where definite speed steps are acceptable).

Starting methods

Starting is achieved using either:

(a) direct-on-line
(b) auto-transformer

Note: Most of the motors start off-load.

Note that the torque produced by a pole change motor depends on the speed that the motor is running at. For the same output power, at low speed the torque will be more, and at high speed the torque will be less.

P.A.M. motor

A P.A.M. motor is a cage-rotor induction motor which uses **Pole Amplitude Modulation** techniques to produce a different number of poles from a single stator winding. One technique employed is to reverse the current in each half of the motor's stator winding to alter the number of poles on the stator and hence the speed of the motor.

Application

P.A.M. machines are widely used where speed is not so critical, as in ventilation fans and some types of pumps, and have made dual-wound stator machines (as fitted to pole-change motors for example) obsolescent.

Remember

A Pole Amplitude Modulated motor offers various speed ratios by pole changing without the need for separate stator windings.

Try this

A contactor control is used to vary the number of poles obtained from the stator windings of a three-phase cage-rotor induction motor. The number of poles obtained by switching are:

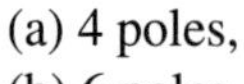

(a) 4 poles,
(b) 6 poles
(c) 8 poles

Calculate the synchronous speed (in r.p.m.) for each connection.

Try this

List 5 applications where variable speed control is required using a.c. three-phase motors.

Single-phase

The single-phase motor used for domestic appliances and power tools is the universal series motor. In Chapter 2 the d.c. series motor was discussed and two methods of speed control shown using variable resistors. In each case the resistors carried comparatively large currents when the motor was loaded. The current produced power losses and resulted in unwanted heat. As pointed out in Chapter 2 speed control in this way can be expensive, bulky and the extra losses reduce the overall efficiency of the motor.

Although a.c. and d.c. series motors are basically the same, when run on a.c. supplies the losses are slightly greater due to the extra reactance of the field windings. Some manufacturers take this into account when designing their speed control circuits and first rectify the a.c. supply.

The circuit shown in Figure 4.57 is typical of some basic electronic ones used on small universal motors. The speed is reduced by controlling the voltage to the motor by using a voltage regulator circuit.

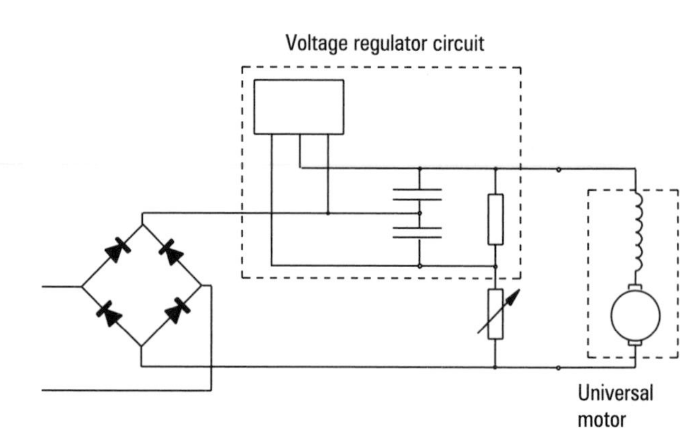

Figure 4.57 *Voltage regulator circuit for speed control of a universal motor*

The regulator is designed so that it has current limiting in the event of overloads. Although this method can be used for small motors the heat produced in the regulator can be a limiting factor. Although the speed is varied when the voltage is reduced the torque is also reduced and there is a limit to the loads that can be used at low speeds. In some cases manufacturers fit larger motors where low speeds are constantly required.

Where larger currents are to be encountered and a greater torque is required at lower speeds, a power controller circuit is used.

These use an a.c. supply in conjunction with thyristors. As a thyristor conducts once it is switched on, until the current returns to zero, the a.c. supply is ideal.

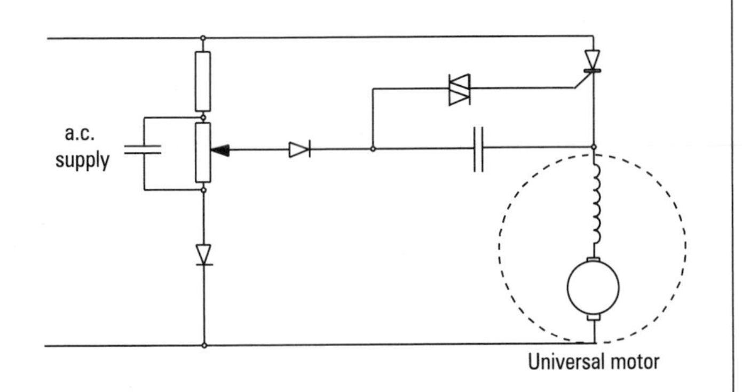

Figure 4.58 *Thyristor controlled circuit*

Figure 4.58 shows a typical circuit where a variable resistor is used to set the voltage that a diac triggers the thyristor to conduct. By adjusting the resistor, the time that it takes the CR circuit to trigger the diac, can be controlled. With this circuit even when the thyristor is fully conducting the motor only runs at about 0.6 of the speed of when it is supplied with full voltage. So that the maximum speed can be used, a switch is incorporated to short out the thyristor. This switch is often fitted so that when the variable resistor is at the end of its travel it automatically comes on giving the maximum speed.

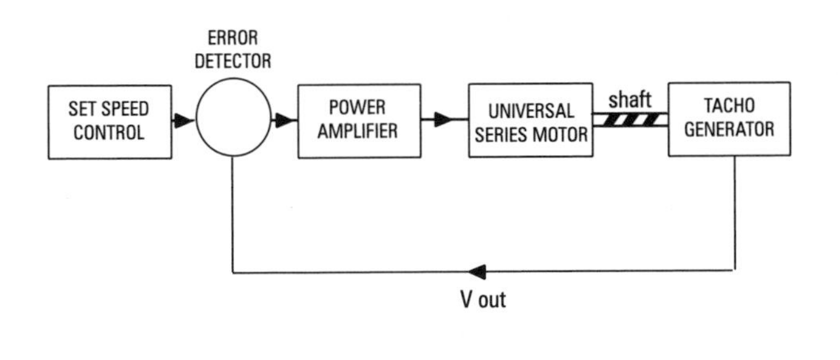

Figure 4.59 *Constant speed arrangement with feed back loop*

To increase the torque at low speed some motors are connected so that a separate supply is maintained on the field winding. Examples of this can be seen in Chapter 2, where d.c. motors are supplied with a.c. supplies.

Where a constant speed is required at different settings a tacho generator is fitted to the motor shaft. This usually consists of a permanent magnet fitted to the shaft and a stationary coil mounted adjacent to it. As the shaft rotates an e.m.f. is induced into the coil by the rotating magnet. The voltage induced into the coil is fed back to an "error detector" that compares it with the voltage set for the required speed of the motor. The controller automatically makes any necessary adjustments to keep the shaft speed at the required value.

Try this

List five applications where variable speed control is required using a.c. single-phase motors.

Exercises

1. (a) Describe THREE methods of starting induction motors that have an output exceeding 12 kW.
   (b) Explain why electric motors have to have a different form of overcurrent protection from other circuits such as lighting and heating.
   (c) State TWO methods of providing overcurrent protection for induction motors and list the advantages and disadvantages of each.

2. (a) Explain the following terms when related to induction motors
   i) synchronous speed
   ii) rotor speed
   iii) slip speed
   (b) A 4 pole, 400 V, 50 Hz, three-phase induction motor, running at 22.9 rev/sec has a full load output of 115 kW. The input current is 225 A at a power factor of 0.83 lagging. For full load, calculate the
   i) input kVA
   ii) input kW
   iii) efficiency of the motor
   iv) synchronous speed
   v) percentage slip

3. (a) Draw a circuit diagram of a three-phase slip ring induction motor include the starting arrangements for
   i) the stator
   ii) the rotor
   (b) Draw the circuit diagram of a direct-on-line starter for a three-phase cage induction motor. Include
   i) overcurrent protection
   ii) undervoltage protection
   iii) local stop and start pushes
   iv) two extra remote start pushes
   v) three emergency stop pushes

4. (a) Compare the following methods of starting a large three-phase cage induction motor
   i) star-delta
   ii) autotransformer
   (b) Draw the circuit diagram of a hand operated, star-delta starter connected to a cage induction motor, and two remote stop units in the diagram.

5

A.C. Generators (Alternators)

Complete the following to remind yourself of some important facts that you should remember from the previous chapter.

Unlike d.c. motors, a.c. ________ motors do not have __________ connections to the ____ parts of the machine.

The speed of an a.c. induction motor depends on the ______ of the supply and the __________ ________ _________ in the motor.

Copper losses occur in the_________ and ________ windings and iron losses occur in the ________ and ________ ______________.

Worn bearings cause___________ losses.

Oil ______ ______ are fitted to ______ type overloads to prevent them tripping when the motor is drawing a high __________ current.

The inch button has ____ contacts, and the normally closed one is connected in series with the ______ ______ contact to prevent the motor running when the button is __________.

Wound rotors have three __________ connected windings brought out to __________ for connection to the ________ ________.

Soft start controllers not only ______ the ________ current when the motor is first switched on, but also control __________ acceleration and ________ build up to meet the requirements of the load.

On a capacitor start–induction run single-phase motor the ______ switch cuts out the __________ winding and the __________ at about ________ full speed.

PAM stands for ________ ________ __________

PWM stands for ________ ________ __________

On completion of this chapter you should be able to:

- ◆ recognise the basic theory of generating a.c.
- ◆ describe the basic construction of an alternator
- ◆ identify the various methods of construction of alternators
- ◆ describe the action of the brushless alternator

In Chapter 1 the basic theory of generation was shown by rotating a coil inside a magnetic field. In practice an alternator is made so that the magnetic field rotates inside the coil. Alternators can be found in many different types and sizes, from cycle "dynamos" to power stations. There are also alternators used in cars, as temporary supplies on construction sites and as stand-by supplies in shops and factories.

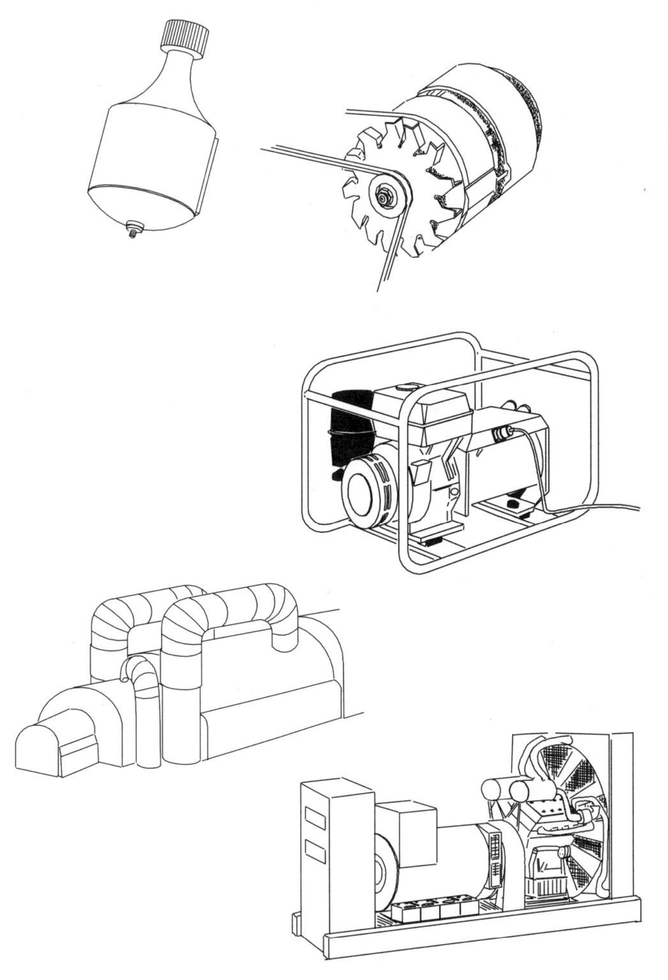

Figure 5.1 Examples of alternators

A cycle "dynamo" is an ideal example of how the basic theory is applied in practice. The magnetic field is supplied from a permanent magnet and the supply is taken from a single coil, as shown in Figure 5.6.

If the coil was rotated, slip-rings would be required to take the generated e.m.f. from the rotor to the loads. By rotating the permanent magnet maintenance of the slip-rings and brushes has been eliminated.

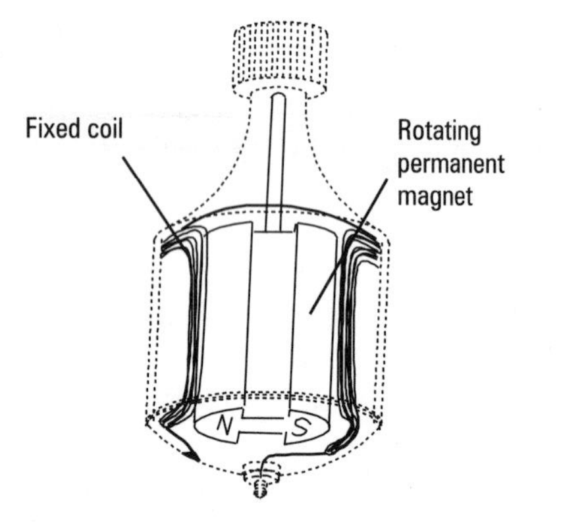

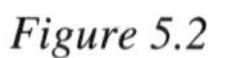
Figure 5.2 *Cycle dynamo internal workings*

On larger alternators the permanent magnet is replaced by an electromagnet. As shown in Figure 5.3.

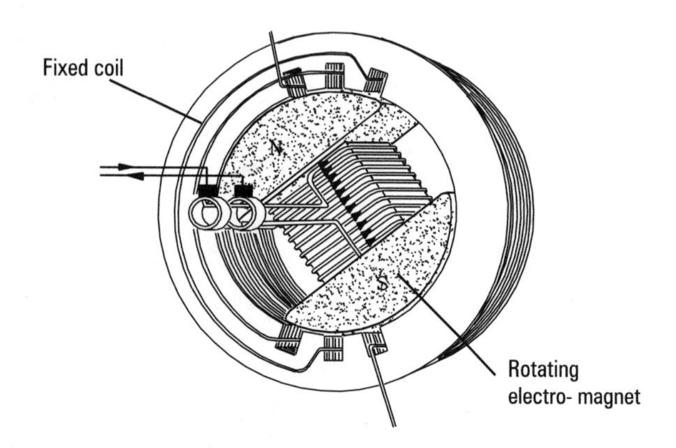

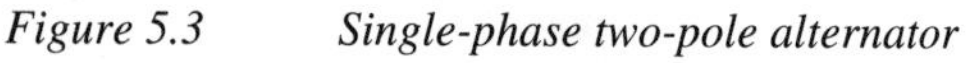
Figure 5.3 *Single-phase two-pole alternator*

It is important to again look at why the rotor is the magnetic field and not the generating windings. If the alternator is to supply a load of say only 10 kVA, 400 V, three-phase, the line current would be

$$kVA = \sqrt{3} \times I_L \times V_L$$

$$I_L = \frac{\text{kVA}}{\sqrt{3} \times V_L}$$

$$I_L = \frac{10000}{1.73 \times 400}$$

$$= 14.45 \text{ A}$$

The current flowing in the magnetic field winding of this type of alternator would be approximately 3 amps.

If the generator winding was turned in a stationary magnetic field, the moving windings would need to be capable of carrying up to 15A and there would also need to be at least three slip-rings and brushes capable of carrying the full load. A fourth slip-ring would be required if the windings were connected in star and the star point had to be connected to earth. By rotating the magnetic field and keeping the generating windings stationary only two slip-rings are required and the current required in the rotating winding is much less.

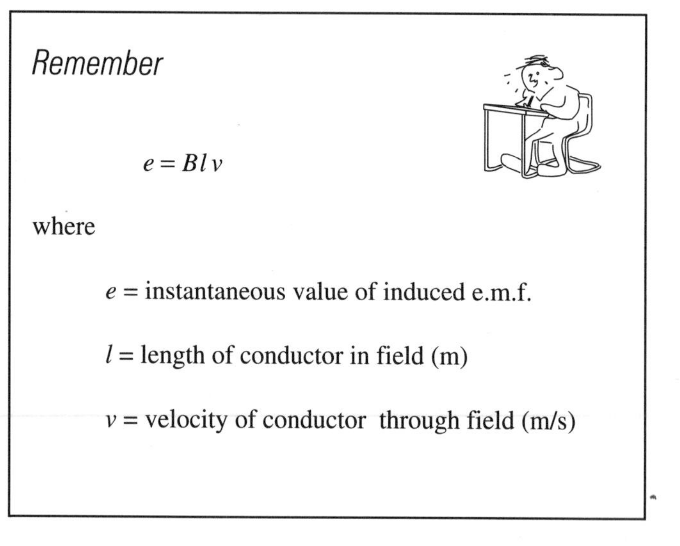

Remember

$$e = Blv$$

where

e = instantaneous value of induced e.m.f.

l = length of conductor in field (m)

v = velocity of conductor through field (m/s)

A good example of an alternator that is made up in this way is that used in a car. The supply to the magnetic field is taken from the battery so no other equipment is required. As the output of the alternator has to be used to charge the battery and supply other d.c. equipment, the output has to be rectified.

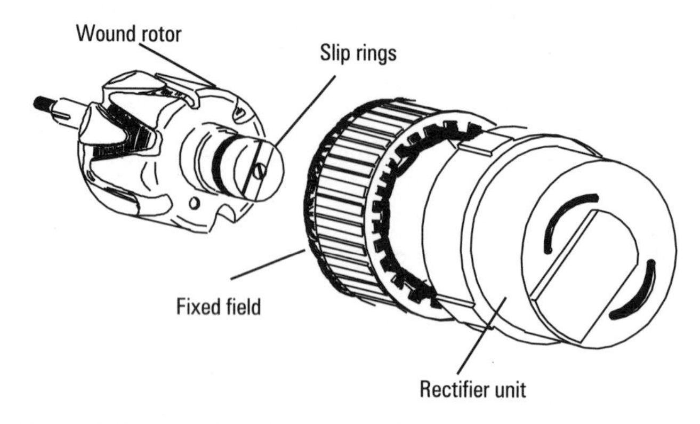

Figure 5.4 *The make up of a car alternator*

To create a smoother d.c. output a multi-phase alternator is used with the appropriate number of diodes.

To create the multi-phases a number of windings are placed in the stator, at regular intervals.

The frequency generated from this type of alternator is

$$f = pn$$

where

f = frequency in cycles per second (Hz)

p = number of pairs of poles

n = revolutions per second that the alternator is being driven at.

Try this

What is the frequency of output of a 8-pole single-phase alternator that is driven at 750 rev/min?

The rotor

Remember that the rotor in an alternator is the magnetic field so the number of poles is taken from the rotor. An alternator that is driven at 1500 rev/min if it is to produce a frequency of 50 Hz must have

if $f = pn$

then $p = \frac{f}{n}$

$= \frac{50 \times 60}{1500}$

$=$ 2 pairs of poles or 4 poles

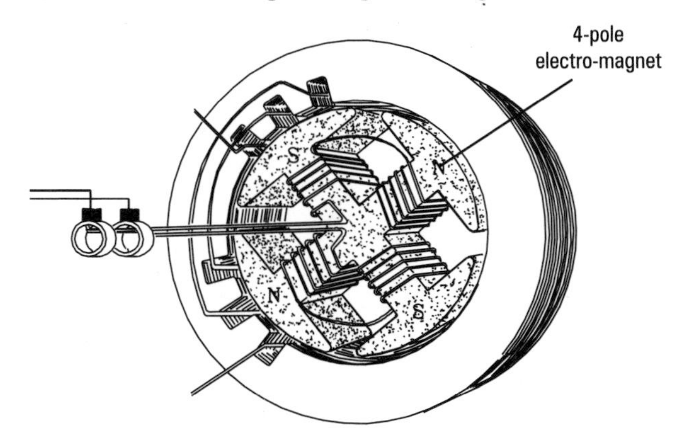

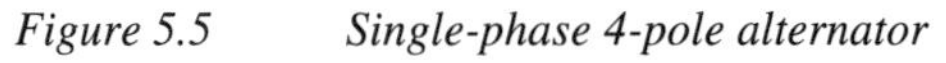
Figure 5.5 Single-phase 4-pole alternator

The rotor has to be designed so that it has a large enough diameter to contain the appropriate number of poles and have the length to create the correct length of conductor.

There are basically two types of construction of rotors
(i) the salient pole type

(ii) those with cylindrical rotors

Salient pole rotors

These tend to be used on relatively low speed small machines where a larger number of poles are required. They can be compared to the field pole of a d.c. motor but the pole face is shaped the opposite way.

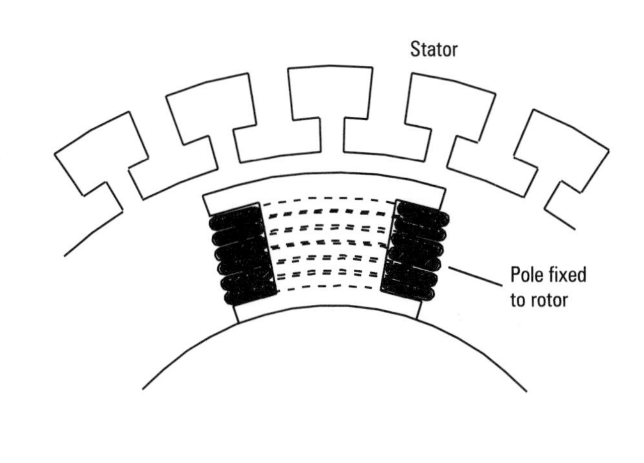

Figure 5.6 Salient pole rotor

Cylindrical rotor

Where rotors are used on high speed alternators they tend to be constructed in this way to help to overcome the high centrifugal forces that are produced. The rotor is usually made of solid steel with slots cut along the length. The conductors are wound to create the poles by spreading them out across the surface of the rotor, as shown in Figure 5.7.

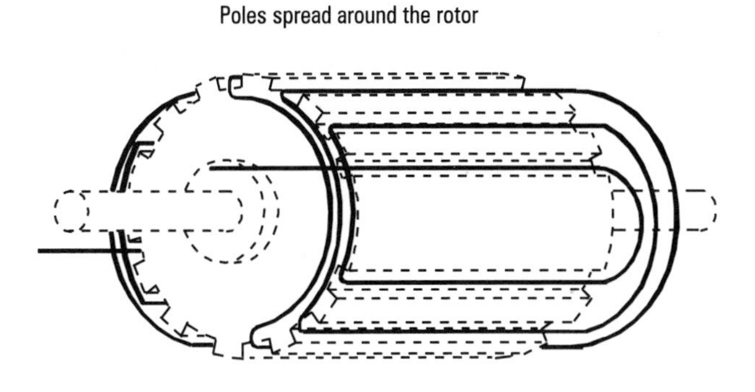

Figure 5.7 Cylindrical rotor

Stator

The windings on the stator are arranged in slots formed in a laminated iron core. The number of coils and the way that they are connected depends on the arrangement of phases and the voltage to be generated. Whereas a single-phase alternator in theory has only one coil a three-phase one has three. Figure 5.8 shows how the three coils are spaced around the stator.

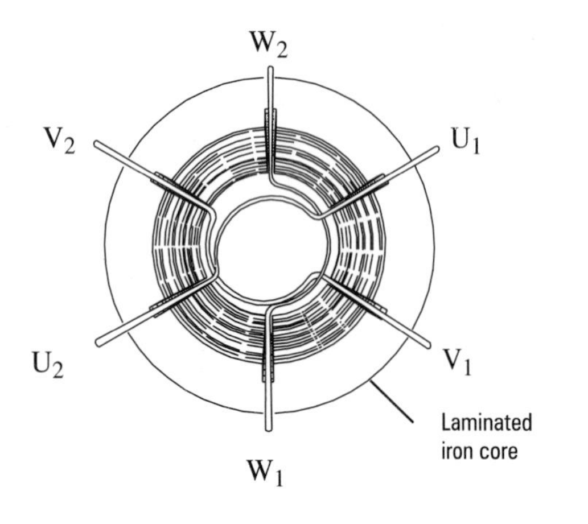

In practice they are more complex than this and Figure 5.9 gives a more practical arrangement.

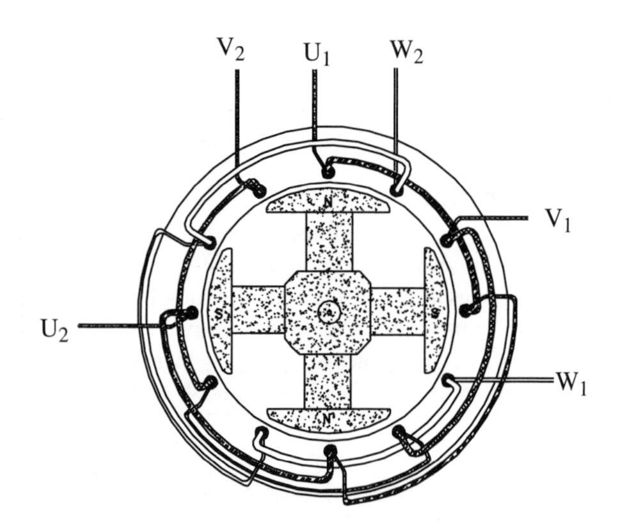

Figure 5.9 3-phase stator with 4 pole rotor

Try this

Sketch the essential parts of a salient pole alternator.

Up until now it has been assumed that where a supply is required for the rotor this is taken from a battery or some other external d.c. supply. On large alternators the d.c. supply used to excite the rotor is generated from a unit mounted on the same shaft as the rotor, as shown in Figure 5.10.

As the shaft to the alternator is rotated, the armature of the d.c. generator also turns and generates a supply. This is connected back to the slip-rings of the alternator rotor to excite the windings. This arrangement eliminates the need for an external supply.

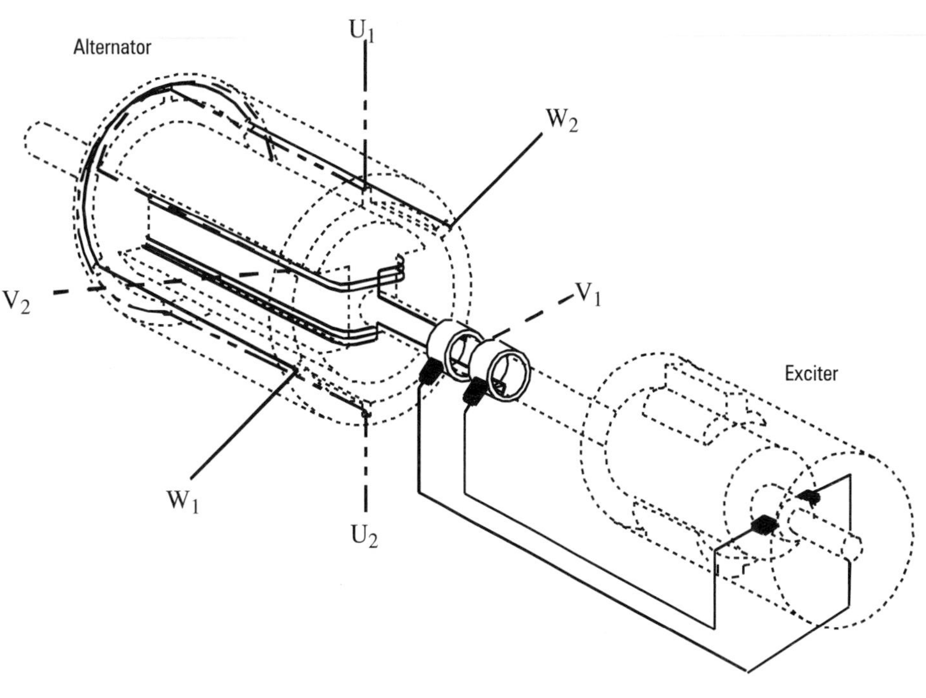

Figure 5.10 *Alternator with external exciter (d.c. generator)*

Brushless alternator

Alternators have now been developed that have no brushes as the generator exciter windings are built into the same rotor as the main windings. This generates a.c. into the coils which is rectified using solid state devices. The d.c. is then taken directly into the main exciter windings. An example of the circuit for this can be seen in Figure 5.11.

The brushless alternator is basically in three parts

- small three-phase generator (exciter generator)
- three-phase rectifier
- main three-phase alternator

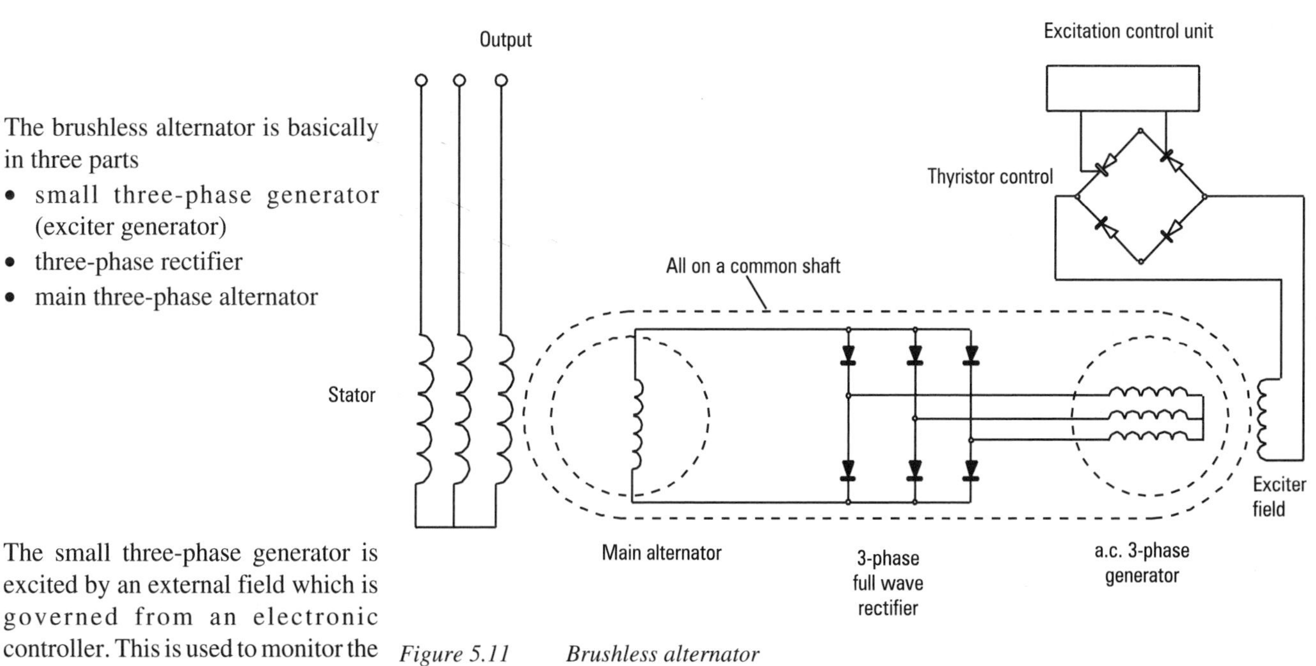

Figure 5.11 *Brushless alternator*

The small three-phase generator is excited by an external field which is governed from an electronic controller. This is used to monitor the main output and automatically adjust the current produced in the exciter generator.

The three-phase rectifier unit is supplied from the exciter generator and feeds d.c. into the rotor of the main alternator.

The main alternator is constructed as an alternator similar to that shown in Figure 5.10. As the exciter and main alternator are mounted on the same shaft with no moving parts between them the efficiency of the overall machine is greater. It also means that by controlling the relatively small current to the exciter generator, the output to the main alternator can be adjusted.

Exercises

1. (a) Describe with the aid of sketches the pole arrangements of an a.c. generator for
   i) salient pole rotor
   ii) a cylindrical rotor
   (b) An a.c. generator has 8 poles and is required to give an output of 230 V at 50 Hz. Calculate the speed at which it should be driven.

2. (a) State the effects on the output of an a.c. generator if
   i) the speed of the prime mover is increased
   ii) the excitation current is decreased
   (b) Explain what is meant by a "brushless" alternator.

6

Synchronous Machines

Complete the following to remind yourself of some important facts that you should remember from the previous chapter.

The magnetic field in a two-pole single-phase alternator is set up by a ____________ ______________.

One advantage of rotating the magnetic field and keeping the generating windings __________ is that only two ______ are required. Another advantage is that the current required in the rotating winding is much ________.

There are two basic types of rotor construction for alternators, they are the ________ ________ type and the ______ type.

Salient pole rotors are used on relatively ________ speed ______ machines where a ______ number of poles are required, whereas cylindrical rotors are used on ________ __________ alternators.

The stator windings of the alternators are arranged in ______ formed in a________ iron ____________.

On large alternators the d.c. ________ used to excite the __________ is generated by means of a d.c. ________ mounted on the same __________ as the alternator's rotor.

The brushless alternator is basically in three parts:
1. a small __________ generator
2. three-phase __________
3. ________ three-phase ________

On completion of this chapter you should be able to:

- describe the basic construction and operation of a synchronous motor
  i) three-phase
  ii) single-phase
- explain the use of synchronous motors for power factor correction
- describe the action of stepper motors
- describe the process of synchronising synchronous generators

The general term synchronous machines can be used to cover both motors and generators, large and small. In this section different types of synchronous motors will be dealt with and some aspects of the synchronising of generators.

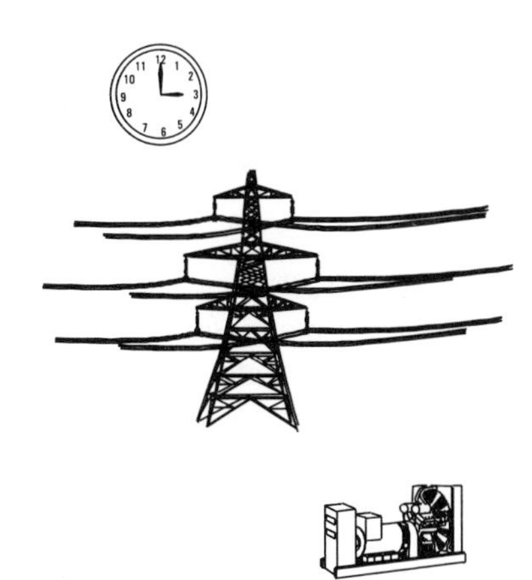

Figure 6.1

Synchronous motors

A synchronous motor is basically an a.c. motor that has a magnetic rotor. The rotor may be a permanent magnetic arrangement or a field supplied with d.c. to create an electromagnet. When the magnet is spinning it locks into the rotating magnetic field of the stator and revolves at synchronous speed. As there is no natural torque developed when the motor is at stand still, methods have to be employed to get the motor started.

Three-phase synchronous motors

A three-phase synchronous motor has a stator winding similar to that of a standard induction motor. The rotor windings usually form one of two types. These are either salient-pole or synchronous-induction.

Salient-pole

The salient-pole type is similar in construction to the corresponding type of alternator. The motor is run up from stationary by starting it as an induction motor. This is achieved by having copper bars embedded in the pole faces and shorting them out with rings at each end of the rotor, these are sometimes referred to as "damper" windings. This creates a caged rotor within the windings. After the motor has run up to induction motor speed, the d.c. supply is fed into the rotor windings and the resulting magnetic field locks into the stator rotating field and revolves at synchronous speed. Figure 6.2 shows the basic circuit for this type of motor.

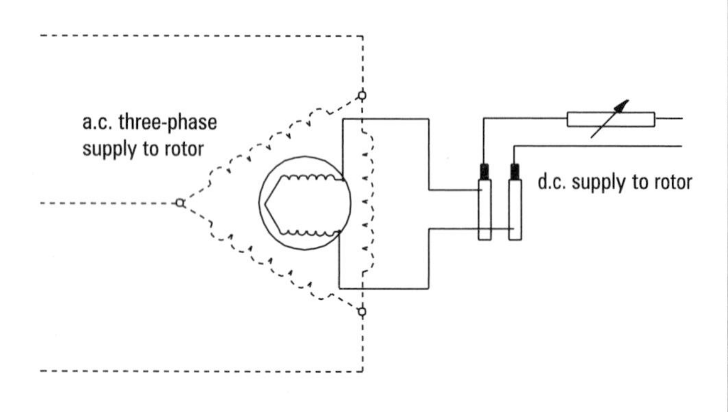

Figure 6.2 *Three-phase motor with d.c. supply to the rotor*

Synchronous induction

The synchronous induction motor also uses a stator similar to a standard three-phase induction motor. The rotor is similar to that of a slip-ring induction motor and is run up to speed using a rotor resistance starter in the same way. When it reaches induction motor speed, a d.c. supply is connected across the rotor windings and the magnetic field locks into synchronous speed. Figures 6.3a and 6.3b illustrate this starting method.

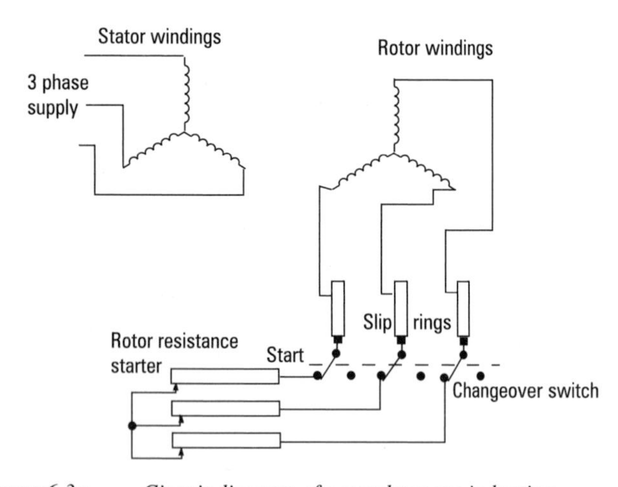

Figure 6.3a *Circuit diagram of a synchronous induction motor – start position*

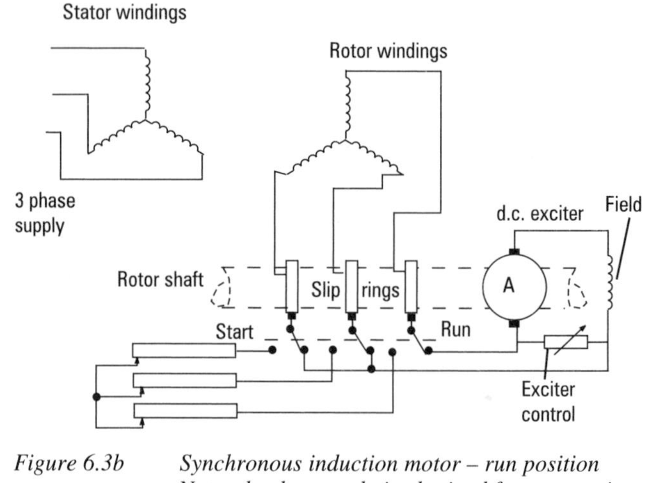

Figure 6.3b *Synchronous induction motor – run position*
Note: the d.c. supply is obtained from an exciter driven off the synchronous motor's rotor shaft..

Synchronous motors are often installed not only to drive some constant load, such as a pump or fan, but also for power factor correction. When the motor is run up to synchronous speed, the d.c. supply creates a magnetic field in the rotor that is similar in strength to that in the stator winding. If the d.c. supply is increased, it excites the rotor so that the magnetising force is greater than that in the stator. This means that the power factor of the motor is increased so that there is no wattless current being supplied to the motor and unity power factor is achieved. If the excitation is increased still further, the stator current can be made to lead the supply voltage giving a leading power factor in the supply cables. This leading power factor can be used to overcome the lagging power factor of other machines taken from the same intake within an installation. Automatic monitoring equipment can be incorporated so that the excitation of the synchronous motor is linked to the overall power factor of an installation. Although this will not correct the power factor of individual circuits, it can be used to overcome penalties being charged by the regional supply company, in the event of a poor overall power factor.

Single-phase synchronous motor

The constant speed of the synchronous motor makes it ideal for applications where time is important. Electric clocks supplied from an a.c. constant frequency source are often driven by single-phase synchronous motors. The construction of these is very basic as they only have one coil which is connected directly across the supply. Figure 6.4 shows a typical construction of a single-phase synchronous motor.

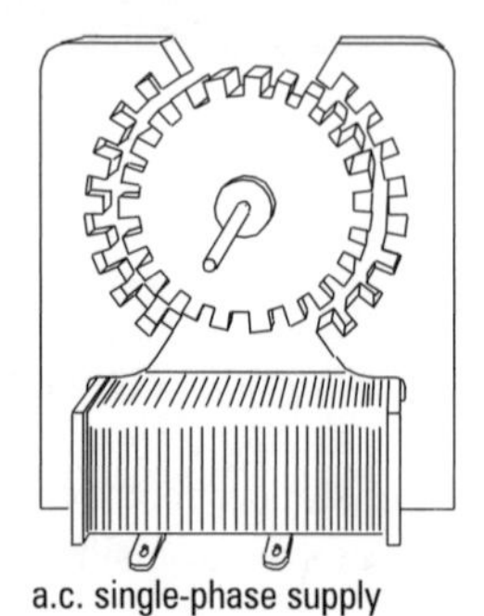

Figure 6.4 *Single-phase synchronous motor*

The stator is the extension of the magnetic field set up in the coil. This has poles cut into it that correspond with those in the rotor. The rotor is made up of a circular permanent magnet with poles cut into it. Like other single-phase motors there is no starting torque but once moving in a direction it will continue to rotate until the supply to the coil is disconnected. The starting of this motor is usually mechanical by an arrangement of springs and pawl and ratchet. It is necessary to ensure that the rotor turns in the required direction as it has no natural direction in itself. The use of this type of motor is very limited as the torque produced is very low.

Remember

The major disadvantage of the basic synchronous motor is that rotor torque is only produced whilst the rotor poles are following the poles of the stator's rotating magnetic field, therefore before torque is produced the rotor must already be running at approximately synchronous speed.

The synchronous motor is started as a wound rotor induction motor via a rotor resistance starter and slip rings. When the motor has run up to speed the switch is changed over (Figure 6.3b), disconnecting the external resistances and connecting the d.c. exciter to the rotor windings. This then causes the rotor's magnetic field to "lock on" to the stator's rotating magnetic field and run at synchronous speed.

Try this

1. Explain the main differences in construction between three-phase cage rotor induction motors and synchronous motors.

2. Explain where a three-phase synchronous motor may be used instead of a three-phase cage rotor motor.

Stepper motors

There are three basic types of stepper motor:

1. permanent magnet
2. variable reluctance
3. hybrid (which is a cross between 1 and 2 above)

Permanent magnet stepper motors are basically synchronous motors. Unlike other motors that are designed to rotate in complete revolutions, the stepper motor is designed to turn through a number of degrees and stop. They are used in many applications that use precise positioning on a rotating shaft. An example is the electronic "daisy wheel" typewriter where the wheel has to spin to the exact position that corresponds to the letter pressed. The stepper motors can be designed so that the angles that rotor moves through are different. For example steps of 6,8,12 or 24 per revolution, are typical. This can be explained simply by looking at Figure 6.5 where a stepper motor has a stator of six poles, 30° apart, and a rotor of four poles. In the position shown the stator poles marked 1a and 1b, are excited.

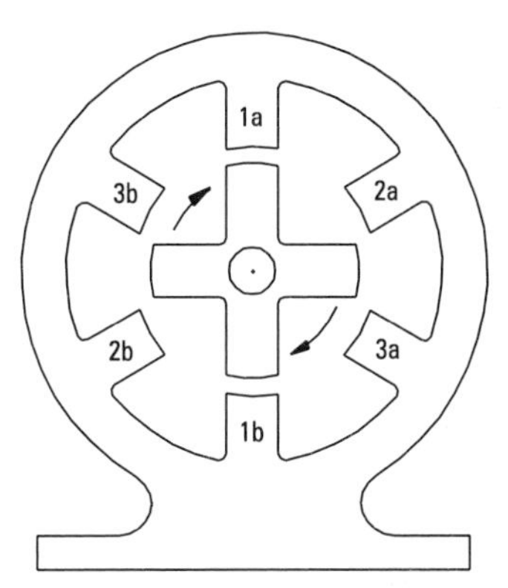

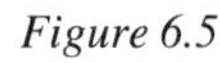

Figure 6.5 *Basic stepper motor principle*

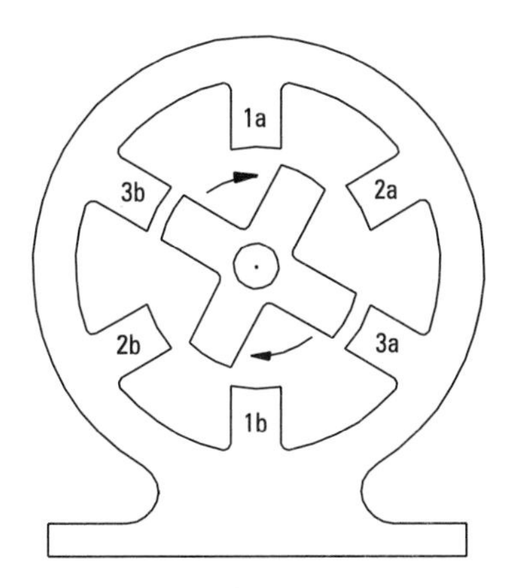

Figure 6.6

To move the rotor from this position to the next clockwise position, the coils around 1a and 1b are switched off, and 3a and 3b excited. The rotor now takes up the new position. By switching different coils on and off, and changing the polarities, it is possible to get the rotor to any of the pole positions around the stator. In practice the motors are more complex and require more positions.

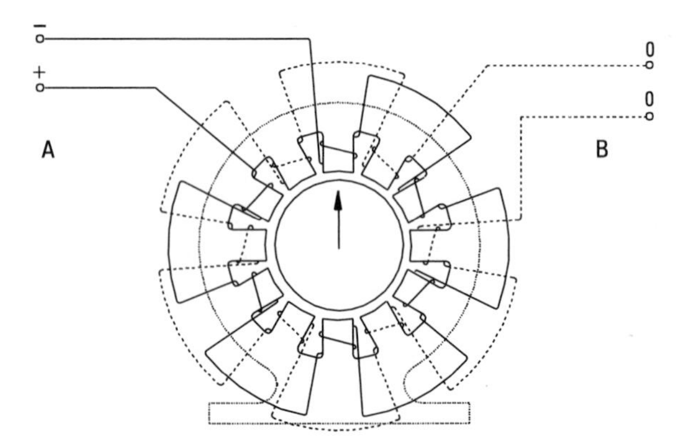

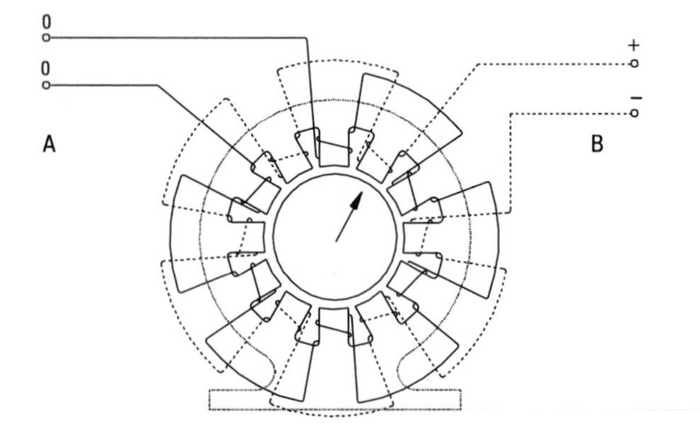

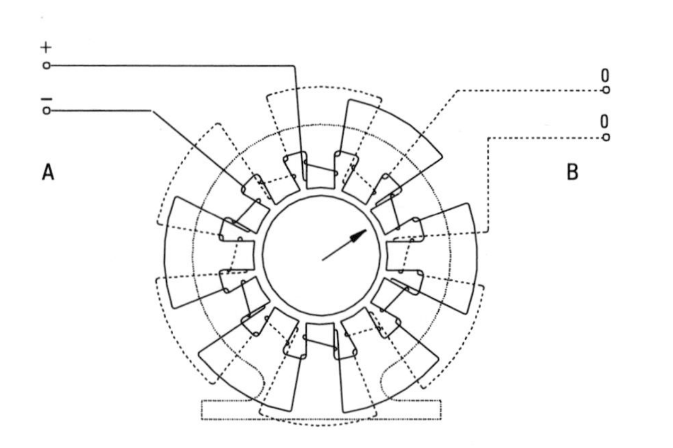

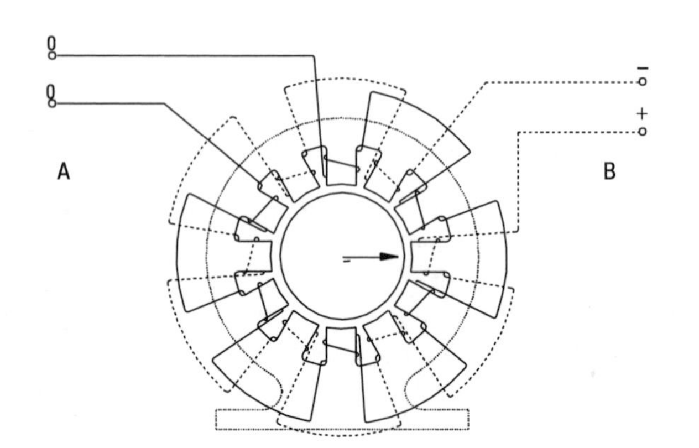

Figure 6.7 *Sequence of a six pole stepper motor*

Figure 6.7 gives another variation where there are six poles connected to two coils. By energising coil A with the polarity as shown, the rotor takes up the first position. When A is disconnected and B connected the rotor moves to position two. Coil B is then switched off and A reconnected but with the opposite polarity. This moves the rotor on to position three. By continually switching the coil supplies in this way, the rotor can be turned through 360° and stopped in any position. There are many practical aspects that need further consideration if an efficient design is to be developed. When it is required to move the rotor, it has to go from stationary to the required position fast. To achieve this a higher voltage is applied, when it has reached the required position only a small voltage is required to keep it there. The control circuitry for this type of motor can be very involved and is usually an integral part of the overall equipment.

Try this

Explain the basic principle of a stepper motor and give an example of where they would be used on:

a) a C.N.C machine
b) a robot
c) a printer

The variable reluctance stepper motor

This type of stepper motor has a soft iron toothed rotor and a stator with projecting teeth. The stator is designed so that it has a greater number of teeth than the rotor.

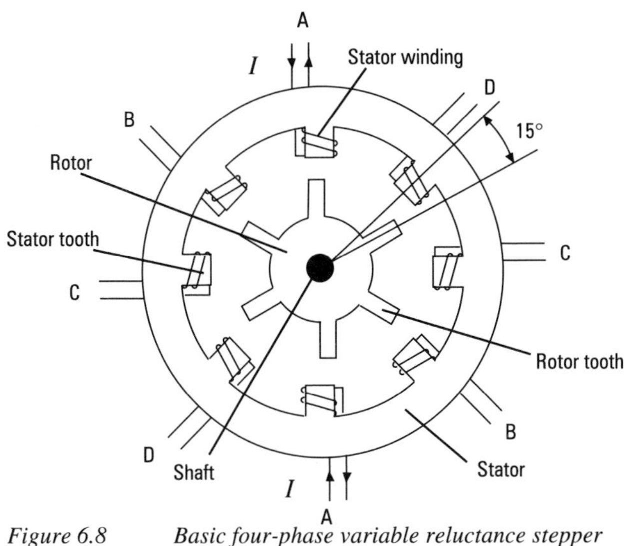

Figure 6.8 *Basic four-phase variable reluctance stepper motor*

In Figure 6.8 the stator has 8 teeth whilst the rotor has only 6 which means that when two teeth are aligned the others are displaced by 15°.

The stator teeth carry four windings A, B, C and D and by sequencing the order in which the coils are energised, the rotor can be made to step from one position to another.

Example

If coils "A" are energised the rotor teeth nearest to the stator coils align with the stator poles "A". If now the d.c. supply is switched to coils "D" the rotor moves through 15° in an anti-clockwise direction to align with the stator poles "D". Further anti-clockwise movement will be achieved by switching to coils "C".

Therefore for anti-clockwise rotation the step sequence is energising stator coils ADCBA or for clockwise rotation ABCDA.

Drive circuits for stepper motor switching

The simplest way of switching the d.c. supply from one set of stator coils to another is to use a standard unipolar drive circuit as shown in Figure 6.9.

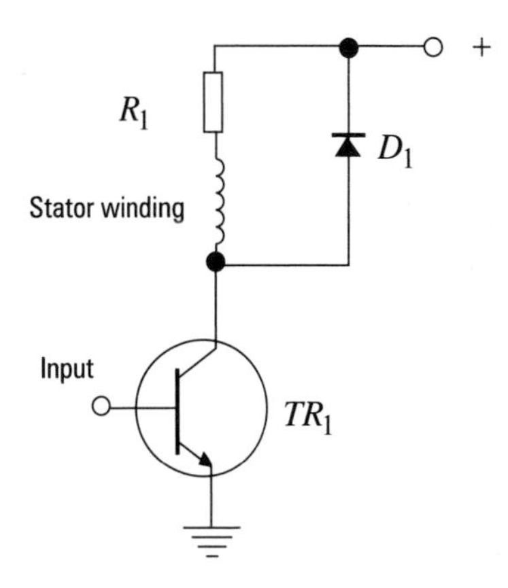

Figure 6.9 *Drive circuit for a single stator winding*

When the supply is switched on to any one of the coils the rated winding current flows, and when it is switched off the diode (D_1) provides a safe path for the induced current to decay without damage to the transistor (TR_1). The added resistance (R_1) reduces the time constant for the winding, enabling the motor to be operated at high switching speeds.

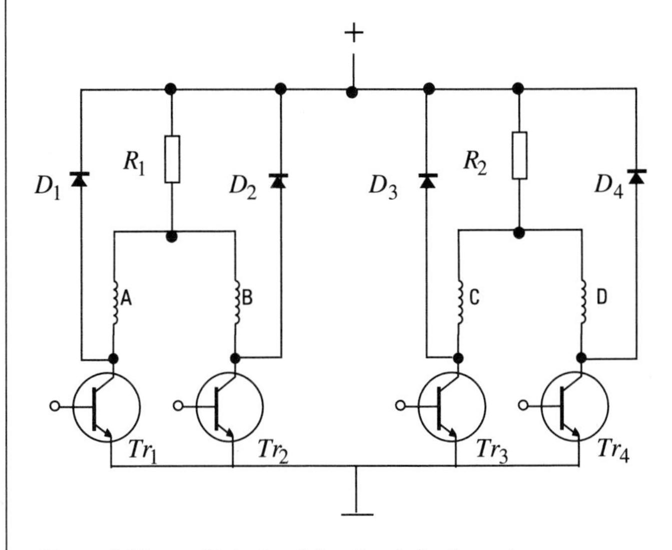

Figure 6.10 *Unipolar drive circuit for four-phase stepper motor*

Remember

A stepper motor is a d.c. motor which is used to convert electronic signals (pulses) into precise mechanical movement.

Step angles can vary from 0.45° to 90°.

Stepper motors provide accurate positional control, and the speed of the motors can be easily controlled by varying the switching (pulse) rates, with the possibility of pulses being delivered as fast as several thousand steps per second.

The rotor torque is very low at high stepping speeds.

Synchronising synchronous generators

All alternators are synchronous machines as they are driven by a prime mover to produce the current at a set speed. When two alternators have to be run in parallel or an alternator has to be connected to an existing supply, there are certain precautions that must be taken.

- the voltages should be equal
- the sequence of the two supplies must be the same
- the frequency of the two supplies must be almost the same
- the voltage across the switch contacts should be very small when the circuit is completed.

  Where large generators are in use, the above precautions would be monitored and switched automatically. On smaller machines it may be necessary to carry this out manually. A simple arrangement that can be used to assist with this process is shown in Figure 6.11.

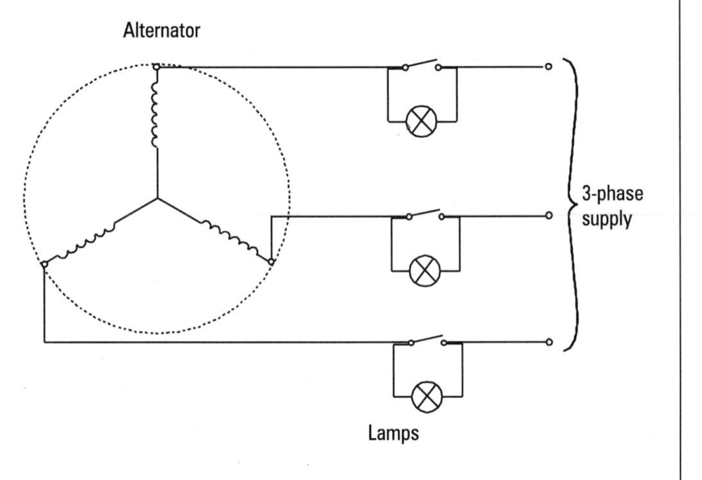

Figure 6.11 Synchronising an alternator to a 3-phase supply

Three lamps are connected across the switch contacts between the two supplies. When one generator is not run up to speed, the three lamps will be lit. As the two loads become closer in voltage and frequency, the lamps will flash on and off. The correct frequency is indicated by the lamps going on and off together. As the generator voltages become closer, the voltage across the switch contacts, decreases. When it is at zero, the voltages are the same from each supply.

Exercises

1. (a) Calculate the speed of a 2 pole synchronous motor working on a 50 Hz 400 V supply.
   (b) Explain why synchronous motors require devices to get them up to speed.

2. A stand by generator is to be run up and connected to a main three-phase supply to supplement the consumer's maximum load. Explain one method that can be used to ensure the generator is synchronised to the supply voltage and frequency before the two supplies are connected together.

7

The Installation of Motors

Complete the following to remind yourself of some important facts that you should remember from the previous chapter.

When the magnetic _______________ is spun it locks into the ________ __________ ____________ of the ________ and revolves at ____________ speed.

The synchronous motor needs to be __________ because it does not produce any __________ from standstill. It may be run up from __________ by starting it as an ________ motor.

The synchronous induction motor has a rotor similar to a ________________ induction motor and is run up to __________ using a ____________ ____________ starter.

Typical applications of a synchronous motor are:
1. to drive a ________ load, such as a __________
2. to improve the ________ ________ of an installation.

The stepper motor is designed to turn through a number of __________ and ____________.

In many applications they are used to precisely __________ a ____________ shaft.

Certain precautions must be taken when two alternators are run in parallel, two of these precautions are:
1. the ____________ should be equal
2. the ______________ of the two supplies must be the ____________.

On completion of this chapter you should be able to:

- discuss the following considerations with regard to the installation of motors
  - electrical
  - mechanical
  - environmental
- describe how to line up a motor with a load to be driven by belts
- select a motor with a suitable enclosure to meet stated environmental conditions
- discuss maintenance requirements for selected motors

When considering the installation of electric motors there are many factors that are going to vary. These can be listed under electrical, mechanical and environmental.

ELECTRICAL

power rating
speed rev/min
supply – voltage
– current
type of overcurrent protection
earth fault protection
installation system
methods of starting

MECHANICAL

type of mounting
drive type
ventilation system
noise control
vibration

ENVIRONMENTAL

temperature
humidity
polluting substances
water and dust
hazardous conditions

Electrical considerations

The first is the power of the machine to be installed. Generally the greater the power the larger the machine. This is the power output of the machine after any losses have been taken into account. Although the actual calculations to determine the rating of specific machines is dealt with in Chapter 8, there are some factors that must be considered now. The electrical supply to a motor must meet the specified requirements.

There are several variables that must be checked, these include the:

voltage
current – a.c. or d.c.

if a.c.

the frequency
single-phase
three-phase

There a number of statutory requirements that must be considered relating to the Electricity At Work Regulations 1989. These include the provision of suitable

- insulation, mechanical protection and correct sizing of conductors
- earthing arrangements
- methods of protection from excess current
- means of isolation from the supply
- environment for carrying out maintenance procedures

Cable systems

When cables are installed to carry the current of motor loads, they must meet the requirements of the appropriate British Standards. This will ensure that the insulation of the cable will be suitable for the voltage range that the motor will be working on. The environment that the cable is to be installed in, will determine the type of cable to be used, and the mechanical protection it requires. Steel conduit systems are an example of where PVC single core cables are used to provide the maximum adaptability on the wiring, with good mechanical protection from the conduit.

When the cross sectional area of a conductor is to be selected for a given motor, the maximum load conditions must be given consideration. Motors when starting can take currents many times greater than their normal running current. The regional supply companies will often determine the maximum rating of a motor that can be switched directly onto their supply. In some cases the high starting currents are only flowing for a short duration and the heat build up in the cables is not sufficient to cause a danger. There are however situations that need special consideration. Some equipment requires motors to start and stop many times an hour with short intervals between the stop and start processes. When this occurs, allowance must be made when determining the rating of the motor cables, to reduce the possibility of heat build up. A factor based on the average possible current, is determined and applied to the input power of the motor. For example the input power of a motor may be calculated at 5 kW but, with regular stop and starting, this may require a factor of 20% to be used to determine the current that the cables are required to carry. In this case it would be

5 kW + 20% = 6 kW

which would need to be divided by the supply voltage to determine the load current.

Another factor that needs careful consideration when calculating the cross-sectional area of conductors, is the voltage drop that is produced. The regional supply companies have the right to drop their supply voltage by a maximum of 6% of what they state it will nominally be. BS7671:1992 allows a further maximum allowance of 4% of the declared nominal voltage, within the installation. These together mean that the voltage at the motor could be 10% below the nominal declared voltage. It must be remembered that a 10% voltage drop will give a 19% loss in power. For this reason the voltage drop should be considered with the manufacturers' data on the relevant motor.

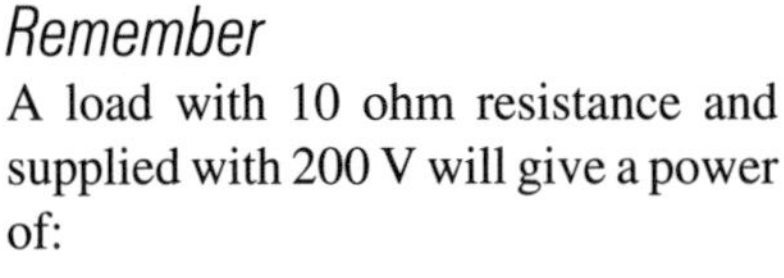
Remember

A load with 10 ohm resistance and supplied with 200 V will give a power of:

$$P = \frac{U^2}{R} = \frac{200^2}{10} = 4000 \text{ W}$$

If the voltage drops by 10% i.e. 20 V

$$P = \frac{180^2}{10} = 3240 \text{ W}$$

Which represents a 19% loss in power.

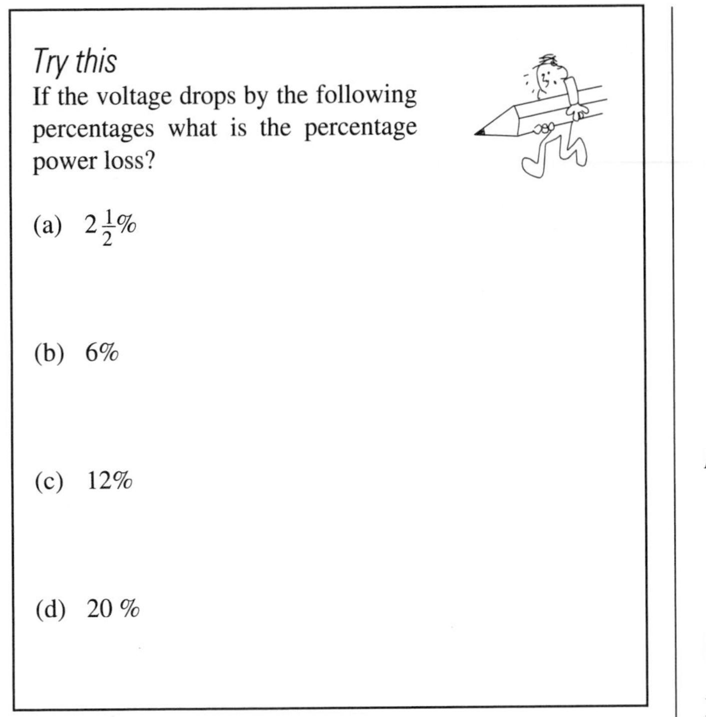

Try this

If the voltage drops by the following percentages what is the percentage power loss?

(a) $2\frac{1}{2}\%$

(b) 6%

(c) 12%

(d) 20 %

Earthing

The general earthing arrangements would be incorporated with the circuit protective conductor of the wiring system. There are however, often problems experienced when making the final connection to the motor due to vibration. These are usually overcome by using some type of flexible arrangement that can withstand the continual movement. Mineral insulated cables, for example, can be formed into a loop and connected directly into the motor terminal box. These have the advantage of combining the circuit protective conductor in the sheath of the cable so no additional conductors are required. Where a number of conductors are required and a conduit system is used, flexible conduit is used for the final connection. Although many types of these are made of metal, none are acceptable as the circuit protective conductor so a separate one has to be installed. Examples of both mineral insulated and flexible conduit connections are shown in Figures 7.1 and 7.2.

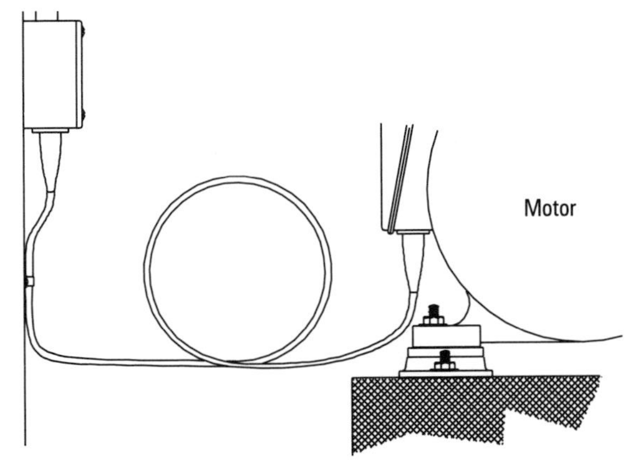

Figure 7.1 *MIMS cable connection to motor*

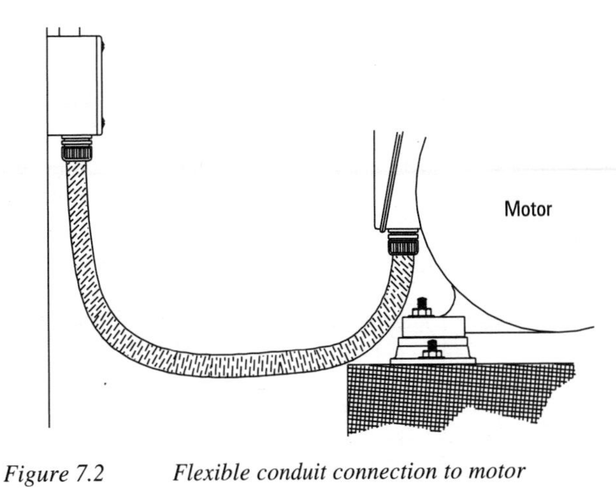

Figure 7.2 *Flexible conduit connection to motor*

Excess current protection

In Chapters 2 and 4, methods of overload protection are examined for incorporation into motor starters. These are designed to cut off the supply to the motor should an overcurrent develop within the motor. In addition to this there must also be over-current protection for the circuit cables up to the motor starter.

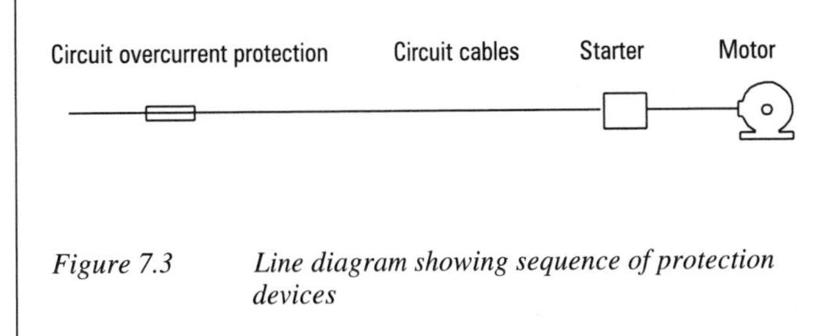

Figure 7.3 *Line diagram showing sequence of protection devices*

As these protection devices have to carry all of the currents to the motor, they have to monitor both the start and the run values. This means that the protection devices have to cope with the starting currents but still be capable of operating under fault conditions. Where the starting currents are up to 600% of the running currents, many overcurrent devices would operate before the motor reached normal working speed. This means that on motor circuits over-current protection devices must be of a type designed with appropriate characteristics. One device that may be considered is the Type 3 or Type C miniature circuit breaker (mcb) manufactured to BSEN 60898 (BS3871). Mcb's have two parts to their operating characteristics, the thermal and the magnetic. Figure 7.4 shows the typical characteristics of a 50A Type 3 mcb.

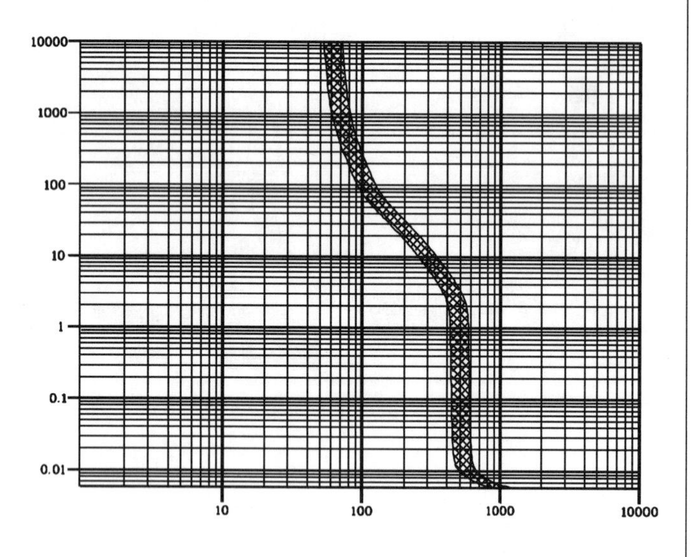

Figure 7.4 *50 A Type 3 mcb characteristics*

The Type 3 mcb has a thermal operating characteristic that takes a longer period of time than the Types 1 or 2.

Depending on the starting characteristics of a motor, this longer period may be long enough to allow the motor to reach a stage where the current is reduced. An example could be a motor that runs on 45 A and is protected by a 50 A type 3 mcb; when starting, the current could reach 600% of the running current. This means that on starting the current could go up to 270 A. From Figure 7.4 the 50 A Type 3 mcb would carry 270 A for just over 10 seconds and this may be enough time for the motor to reach a speed where the current is reducing. Under fault conditions where 1000 A could be flowing, the same device would operate within 0.01 of a second. As a guide on a.c., control gear manufacturers suggest for a 3 phase motor on 400 V a factor of 8 times full load current will give a peak starting current (stalled motor current), a starting characteristic of 6 times full load current for up to 1 second should be used. This applies to motors started direct on line or by star-delta starting, since the inrush transient at changeover in star-delta may be at least as great as the D.O.L. peak inrush.

Miniature circuit breakers do have limitations and many motor manufacturers prefer to specify High Breaking Capacity (HBC) or High Rupturing Capacity (HRC) fuses to offer backup protection in motor circuits. Motor rated fuselinks (Gm) to BS88 are designed to meet the requirements of motor starting. The fusing factor for these was known as Q1 type but now that BS88 has aligned with the international fuse specification IEC 269 the concept of "fusing factor" has been replaced with "utilization category" and the Gm category has been introduced. As with m.c.b's manufacturers have a guide they work to. For HBC fuses it is based on the assumption that the starting conditions for typical 3 phase 4 pole 400 V motors are 8 times full load current for 6 seconds when connected direct on line. When started in star delta 4 times full load current for 12 seconds, is used.

Isolation

The Electricity At Work Regulations require that circuits are not only isolated but that this is carried out in a way that ensures that the isolation is secure. This means that once the isolation has been carried out it is not possible to be "accidentally" switched back on again. In many cases there may be a requirement for two levels of isolation, one for the circuit as a whole, the other for the motor only.

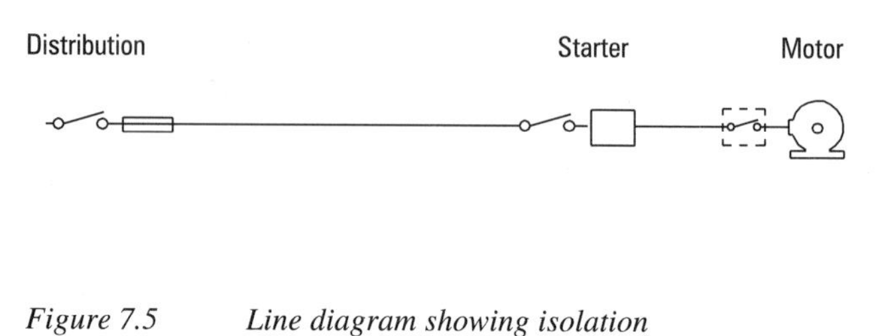

Figure 7.5 *Line diagram showing isolation*

The circuit as a whole may be isolated at a distribution board but it must be remembered that it is a requirement to securely isolate. There must be a method incorporated to ensure the circuit is locked off and can not be re-instated until required. Isolation for the motor depends on its position relative to the circuit isolation. If the control gear and motor are remote from the distribution these will require separate secure isolation. After isolation has been carried out and before any work is carried out, tests should always be made to ensure that the isolation has been effective.

Maintenance

When machines are installed, maintenance must be given consideration along with all of the other factors. Often machines have been sited in positions that are ideal for their working but a nightmare for the maintenance staff. It is a requirement however that there is adequate working space, access and lighting, to carry out work without danger. This means that adequate space, access and lighting for maintenance, should be built in at the same time as the machine.

Mechanical considerations

The mechanical variations may be considerably different from motors to generators. Usually generators either come complete connected to the prime-mover, or they are installed by specialists. Motors are often an integral part of machines but sometimes have to be installed separately. In any case at some stage decisions have to be made with regard to conditions such as the:

- mounting type and methods of drive
- methods of ventilation
- environmental protection

Assuming all the electrical considerations have been given and a motor has been selected to meet the load requirements, the mechanical factors must be considered. These are often inter-related and a decision made on one will directly effect another.

Mounting type and methods of drive

The method used for fixing the motor to a solid surface, varies with the application and often the drive type. Foot mounted machines are probably the most widely used. These consist of four flat flanges fitted along the base of the machine, as shown in Figure 7.6.

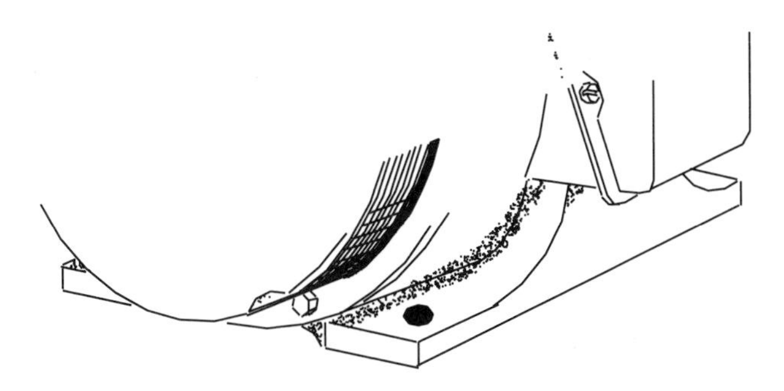

Figure 7.6 Foot mounted motor

The feet are bolted to a solid base which may allow for adjustment to align drive shafts. Slide rails are fitted to the solid base and the motor is bolted to them.

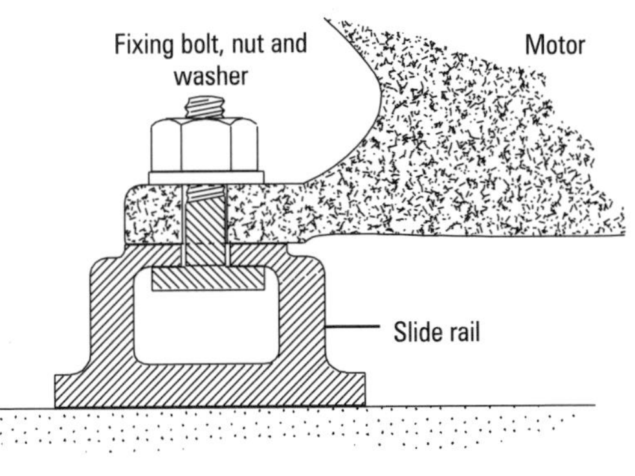

Figure 7.7 Section across slide rail and motor foot

The motor is fitted to the slide rails so that the belt tension can be adjusted.

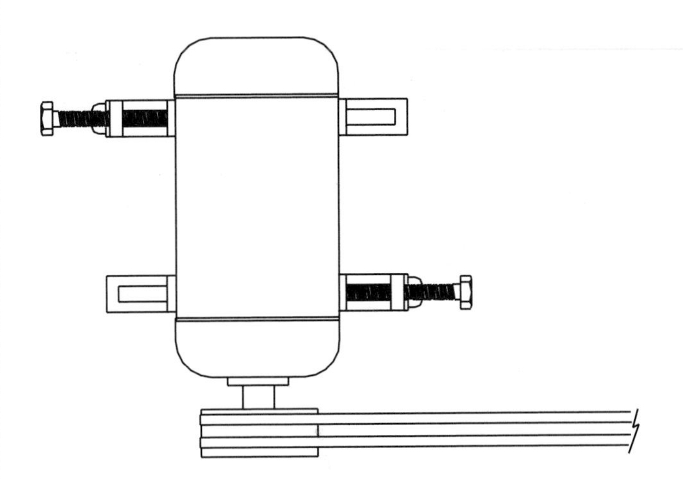

Figure 7.8 Position of slide rails

The alignment of belts is carried out by adjusting the position of the pulleys on the drive shafts. The slide rails need to be fitted to solid material so that they can not become loose with vibration. When a motor is an integral part of complex plant, the frame of the plant is often used for mounting the slide rails. This can be designed by the manufacturer to meet all the necessary requirements. Where a motor is to be mounted on the fabric of a building, the slide rails are normally fitted to a concrete slab built for the purpose. In this case the electrical engineer would work closely with the construction staff to ensure bolts were sited to line up with the slide rail fixings. The bolts used for this would be the rag bolt type, so that a good key with the concrete was obtained. After the concrete had completely set the rails and motor could be fitted.

Table 7.1 shows examples of the depth of concrete foundations.

Table 7.1

Rating of motor	Depth of concrete
up to 9 kW	10 to 13 cm
9 kW to 22 kW	13 to 25 cm
22 kW to 37 kW	25 to 30 cm

Where a motor is to be fitted to an existing concrete slab and no bolts are fitted, the slab can be drilled and expanding bolts used to secure the slide rails.

To correctly align the motor to a load, when belts or chains are being used, the shafts of each must first be parallel. They must also overlap so that the two sets of pulleys can line up.

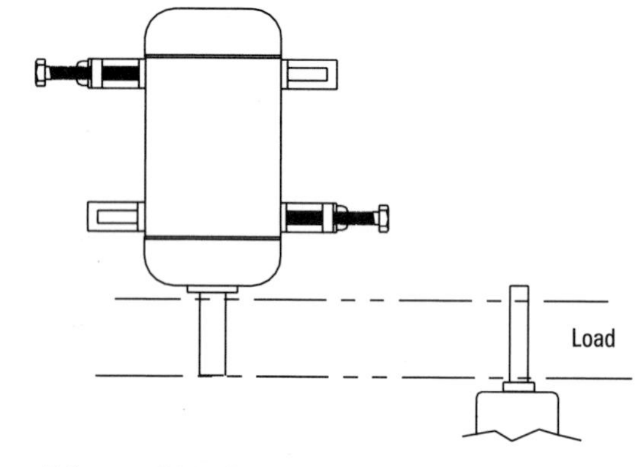

Figure 7.9 Lining up motor and load

The slide rails on the motor give some adjustment to the parallel aspect but the motor can not be moved forward or backwards. To align the pulleys, they should be placed on each shaft and one fastened in an approximate position. Placing a string along the side of the fixed pulley so that it touches both edges, can be used to line up the second pulley. By sliding the loose pulley on the shaft so that the string touches both edges of both pulleys. Where small diameter pulleys are in use a straight edge should be clamped across the face of the pulleys. When both pulleys are in alignment they can be tightened in position. The belts can then be fitted and final adjustments made.

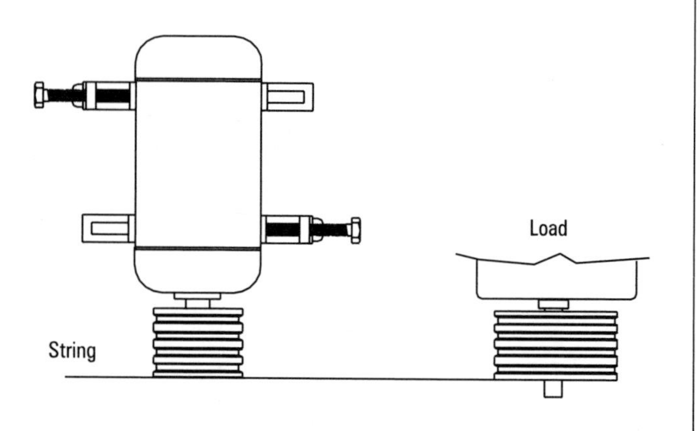

Figure 7.10 Using string to line up motor

In many cases motors have to be bolted directly onto equipment and are fitted with a flange for this purpose. The flange is formed on the front face of the drive end plate. The face of the flange is ground to form a tight straight fit to the load. To ensure the shaft of the motor and the load are exactly in line the fixing holes are drilled to precise measurements. The shafts are connected together by a direct coupling arrangement.

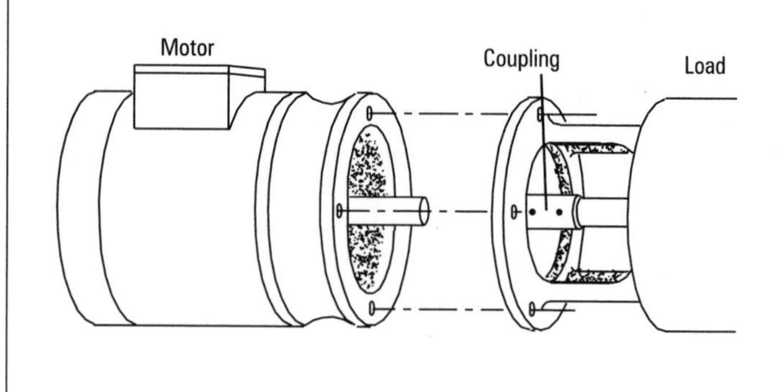

Figure 7.11 Direct coupling

In Figure 7.11 the open skirt is part of the load casing, in some circumstances this is an integral part of the end plate of the motor. It is necessary to have some open arrangement to make off the drive coupling between the motor and the load.

Method of ventilation

There are basically two methods of cooling electric motors,

- one by continually passing air through the windings and iron core
- two allowing the heat from inside the motor pass into the casing and then cool the casing. The method used usually depends on the environment the motor is to be installed.

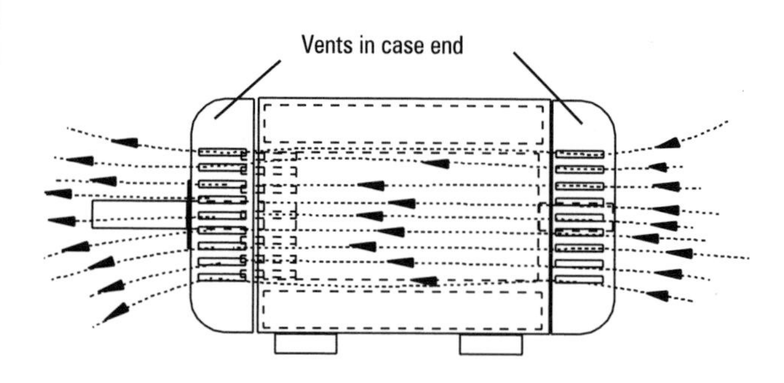

Figure 7.12 Ventilation through the windings

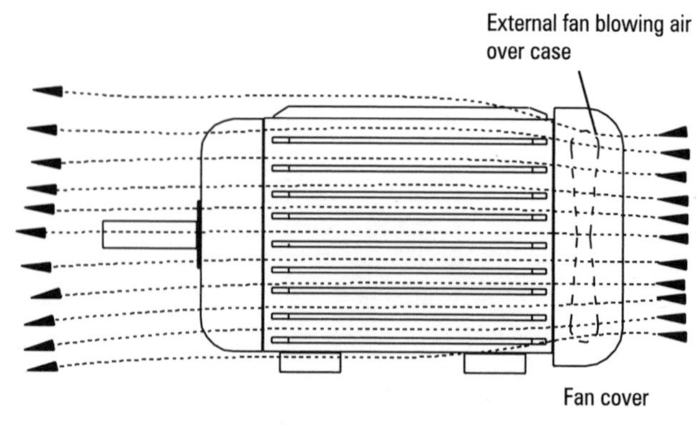

Figure 7.13 Ventilation over the windings

Environmental considerations

As the cooling of motors is either through the windings or over the case, the environmental conditions directly affect the method of ventilation. A motor that uses the full flow of air through the windings is limited to comparatively dry, dust and vapour free, environments. Where there is any hazards present some protection must be given.

In a dry dust and vapour-free environment a motor can have a free flow of air through the windings. Passing in one end plate through the windings and out of the other end plate. The only limiting factors to the flow of air are grills placed in the end plates to stop the entry of foreign bodies. This type of enclosure is known as a screen protected type.

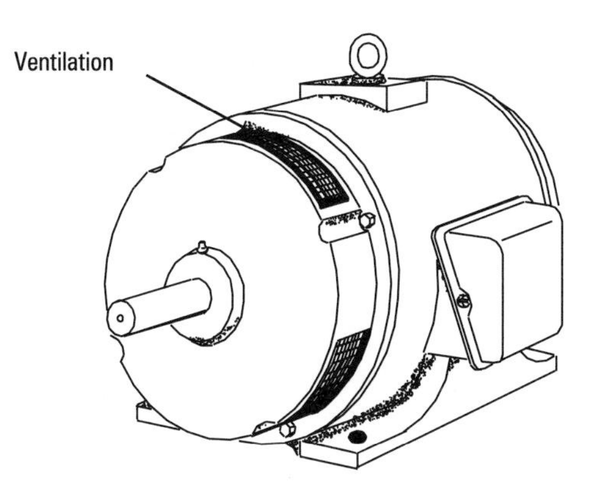

Figure 7.14 *Screen protected enclosure*

Where there is moisture in the air and drips of water may fall onto the motor the end plates are covered at the top. This is known as a drip proof motor enclosure.

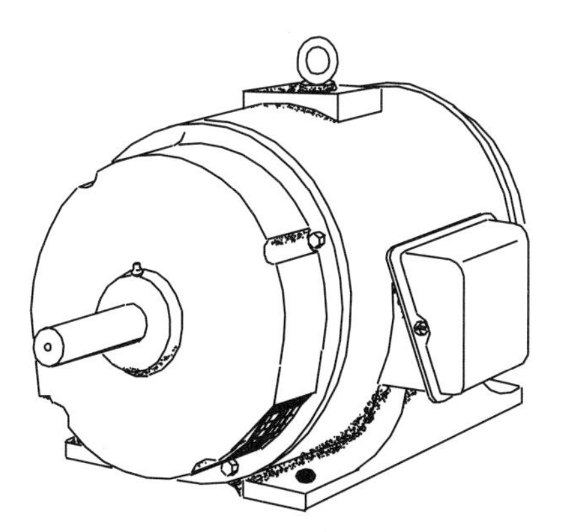

Figure 7.15 *Drip proof enclosure*

As soon as there is any amount of moisture, dust or vapour in the air the motor enclosure must be such that the contaminated air does not pass through it. this may involve using a duct to get clean air to the motor, as in Figure 7.16, or the use of a totally enclosed motor Figure 7.17. This means that no air can pass through the motor so cooling takes place by forcing air over the case and clearing all radiated heat.

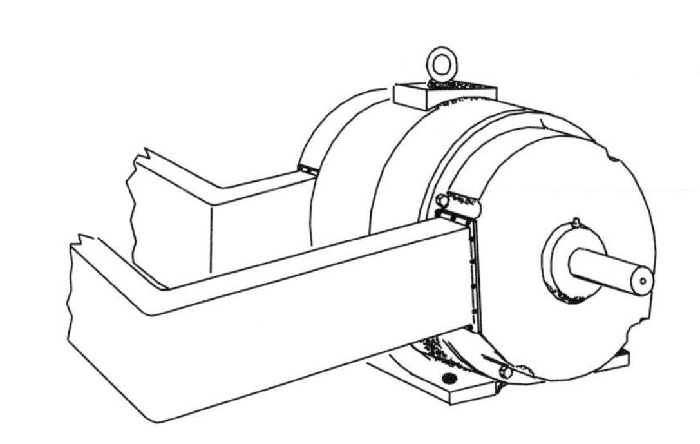

Figure 7.16 *Duct ventilated motor*

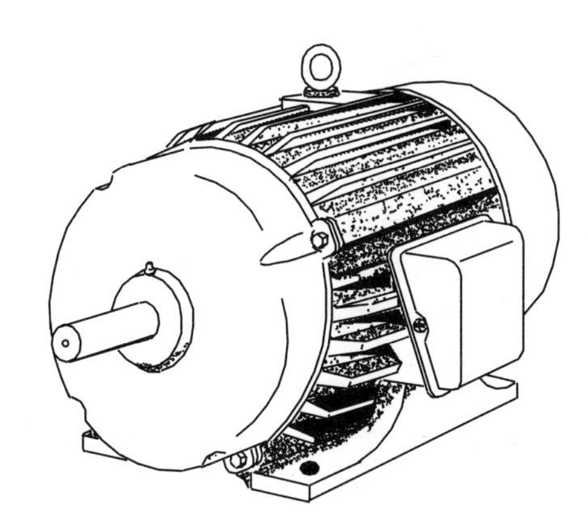

Figure 7.17 *Totally enclosed fan ventilated (TEFV) motor - showing cooling system*

Where a motor is to be installed in an environment that contains hazardous vapours that could explode if ignited, a very heavy form of totally enclosed motor is used. The case is made so that all faces that are bolted together are ground surfaces. The connection box on the motor is usually very large compared with a similarly rated totally enclosed machines.

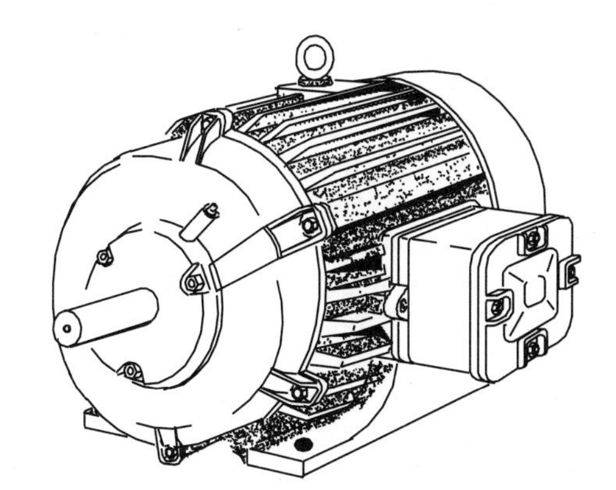

Figure 7.18 *Flameproof motor*

The use of these motors is very specialised and they should only be installed by suitably trained engineers.

Maintenance

In addition to the installation of motors their maintenance must also be given consideration.

Maintenance of motors can be divided into three parts:
- the motor itself
- the control gear
- the drive system

The motor maintenance depends on the type of motor in use. D.C. machines that have commutators and brush gear often require more maintenance than other motors. As the brushes are in mechanical contact with the commutator continuous wear takes place. The length of brush needs to be checked against new ones. Where there is a suitable length left consideration must be given to when the next maintenance is scheduled and whether the brush would last that long. Commutators also wear and these need to be checked to ensure the commutator segments are higher than the spacers between them.

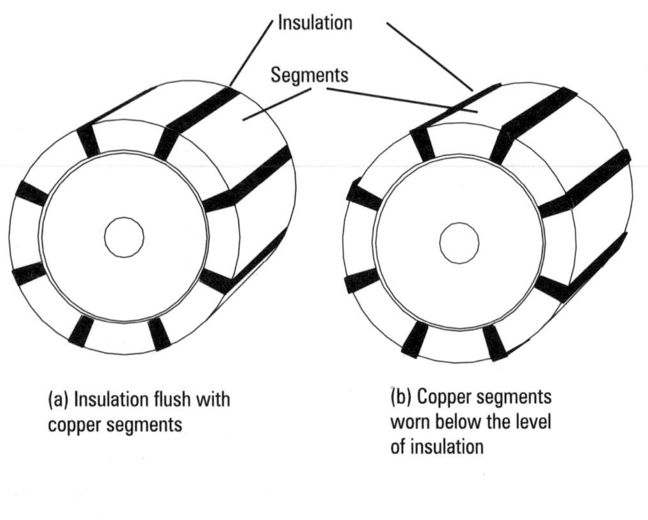

Figure 7.19

If, as in Figure 7.19a, the copper segments are flush with the mica spacers, or in the extreme case as in Figure 7.19b the spacers are higher than the segments, maintenance is required. Assuming the commutator segments are still comparatively thick and worth maintaining, the mica spacers must be cut so that they are once again lower than the copper segments. If this was not carried out excessive sparking would take place due to the poor electrical contact. Where the copper has worn lower than the harder mica, the commutator must first be turned on a lathe to bring it all to the same level. Then the spacer must be cut in such a way that the segments are not damaged. This is usually sawn using a fine blade in a holder and patiently cutting each spacer independently, as in Figure 7.20.

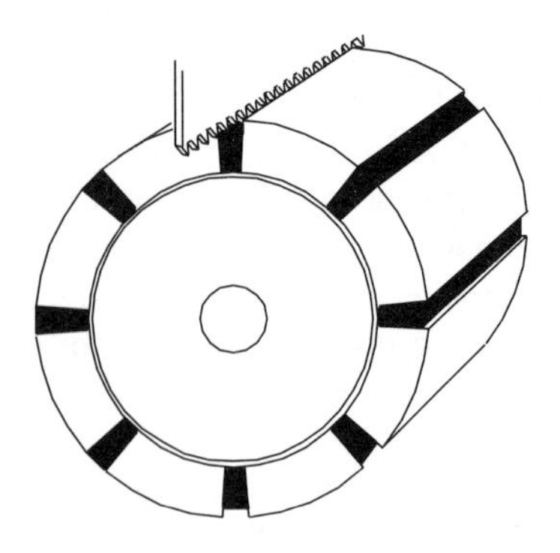

Figure 7.20 *Insulation being "undercut" so that it is below the level of the copper segments*

A.C. slip-ring motors also require brush maintenance. As there are no commutator segments to check the maintenance is not so complex. There are however, other factors to consider. The brushes on slip-ring motors often contain a high content of copper. This is transformed into dust as the brushes wear down and should be cleared from the motor when maintenance is carried out.

All motors will gather dust when in use. This can cause cooling problems if left to build up. Maintenance should include clearing out all ventilation ducts and holes. As all motors revolve on bearings these should always be checked. Lubrication varies from motor to motor but manufacturers' information should give advice on the correct type and regularity of checking and if "sealed for life" bearings have been used. Often a simple turn of the motor by hand can highlight the sound of bearing wear.

Remember

The mica insulation is harder than the copper commutator segments.

Control gear

The work that has to be carried out to maintain motor starters and associated control equipment, varies tremendously between types and manufacturers. There are usually three main parts:
- main contacts
- overload
- control circuits

The main contacts for d.c. motors are completely different to those used for a.c. The d.c. face-plate starter has large copper contacts that must be kept clean and level, so that the wiper contacts can pass from one to another. In contrast a.c. contacts can sometimes be cleaned but often have to be replaced when they become burnt. Contacts that have many starts and stops in an hour generally require more regular maintenance than less operated ones, even if the running time amounts to the same.

Overloads require checking not only for contact condition but also correct setting. In some cases where oil filled dashpots are used, the state of the oil needs to be checked, filled and/or replaced.

Control circuits generally are low current circuits carrying only the operating coil current. This means that contact wear is not so great. In some environments, dust and damp can lead to mechanical failure as push buttons clog up and do not work correctly. If this is a serious problem, devices designed for adverse environmental conditions should be installed.

Drive systems

A drive system can be one motor directly connected to a load or one motor belt driving one load. It can also be one motor belt driving a shaft which has a number of loads taken from it.

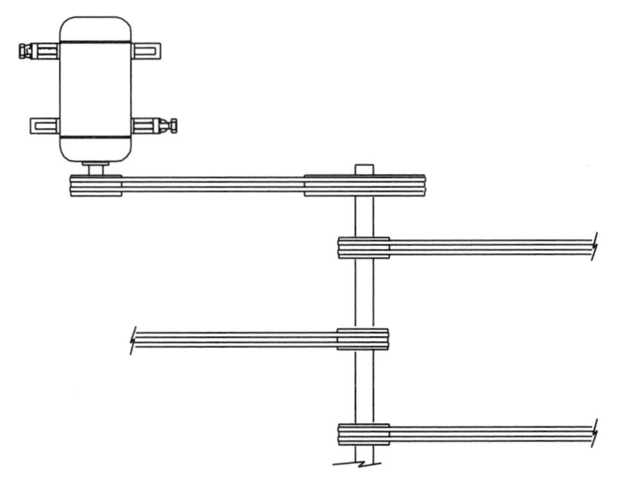

Figure 7.21 *Multiload system from one motor*

There are several advantages of having a single motor running a number of loads. Most of this are related to the fact there is only one motor to start and maintain. There are however disadvantages many related to health and safety problems. Within the working environment all belts and pulleys must be guarded so that they cannot be a danger. With the single drive system some belts become very long and difficult to guard without causing obstructions in the work area. There is also the problem of providing power for all of the loads even if some are not being used. A motor running on light load not only is inefficient but if run on a.c., also has a poor power factor.

Single motors driving a single load cost more initially but have the advantages of being started and stopped when required.

Maintenance on all drive systems means inspection of coupling, or pulleys and belts. Couplings may need to be tightened as vibration can work fastenings loose. Belts should be checked for wear and undue slack. When belts are replaced adjustments need to be slackened off before either the old belt is removed or the new belt fitted. Adjustments should then be made to get the correct tension on the belt. Where multi-belt drives are used, belts should be replaced in sets and not individually. This ensures all belts can be set at the same tension and older belts are not loose while new belts are tight.

Insulation testing

When motors have been out of commission for some time, in addition to the mechanical tests, an electrical insulation test should be carried out. Before any instruments are used the motor should be completely isolated from every supply and control equipment. If any conductors are still connected to the motor after isolation has taken place, a check should be made with a voltage tester to ensure no supplies are still there. The instrument used for insulation resistance testing of a motor is a 500 V d.c. megohmmeter. Tests should be carried out between each winding and the frame of the motor NOT between windings as this would only give the resistance of the conductors. The minimum resistance acceptable between windings and frame is 0.5 OHMS. Care should be taken on small single-phase motors where suppressor units are fitted. These should be disconnected before tests are carried out.

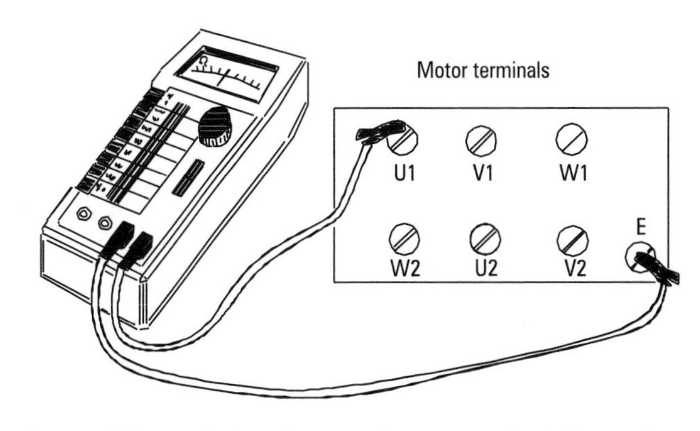

Figure 7.22 *An insulation test between each winding and earth*

Try this

Explain the need for carrying out an insulation resistance test on a cage induction motor that has not been used for some time.

Motor nameplates

We see many types of motor in various shapes and sizes driving machinery.

How can we identify which type it is?

We can identify it by looking at the nameplate which is affixed on the motor frame.

Motor nameplates provide information vital to the proper selection and installation of the motor. Typical information found on motor nameplates is described as follows. (Figure 7.23)

Serial Number is the specific motor identification.

kW is the rated output power of the motor in kilowatts.

R.P.M. means the speed in revolutions per minute when all other nameplate conditions are met.

Volts is the operating voltage of the motor.

F.L.C. is the "Full load current" drawn from the supply line, when the motor is operating at the rated voltage and frequency at the fully rated nameplate power.

Phase indicates the number of voltage phases at which the motor is designed to operate on.

Rating or Duty is the cycle of operation that the motor can safely operate. This motor has a "Continuous" rating, which means it can operate fully loaded 24 hours a day without overheating.

Hz (Hertz) is the frequency of the supply system for which the motor is designed. Performance will be altered if it is operated at other frequencies.

Class is the "Class of insulation", and refers to the insulating material used in winding the motor stator Class E insulation has a 120° C limit.

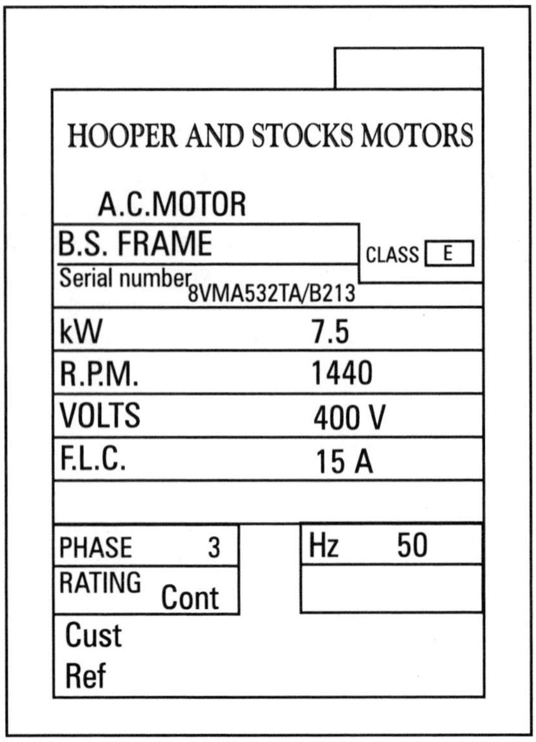

HOOPER AND STOCKS MOTORS
A.C.MOTOR
B.S. FRAME
CLASS E
Serial number 8VMA532TA/B213
kW 7.5
R.P.M. 1440
VOLTS 400 V
F.L.C. 15 A
PHASE 3
Hz 50
RATING Cont
Cust
Ref

Figure 7.23 Motor nameplate

Exercises

1. (a) Describe the procedure for lining up a foot mounted motor on slide rails to a load which is to be driven by four parallel vee belts.
   (b) When a motor is started on load, the belts slip on the pulleys before gaining speed. List the action that should be taken to identify and rectify the problem.

2. (a) Explain two methods of cooling a.c. induction motors.
   (b) Give an example of suitable motor enclosures for each of the following situations. Explain the reasons for each choice.
   i) A motor driving a fan inside a heater unit.
   ii) A motor driving a compressor which is sited outside with only a roof over it.
   iii) A motor driving the pump in a petrol delivery pump.
   iv) A motor driving an extractor fan in a flour mill.

3. A water pump is coupled to a 230 V, 600 W single-phase capacitor-start induction-run motor by a vee belt drive. The motor has run satisfactorily for several months and been out of service for two months. When attempting to restart the motor, it just hummed loudly and would not drive the water pump.
   (a) How would you verify whether the fault is on the motor or on the pump?
   (b) If it is found that the fault is on the motor, how would you check which winding is faulty?
   (c) If the auxiliary (start) winding circuit is faulty, state two typical defects.
   (d) How would you establish which of the defects in (c) is present?

4. A self-contained, diesel driven standby generator is to be installed outside a food processing factory to enable the factory to continue production in the event of a mains supply failure.
   You have the responsibility of advising the factory plant manager on the specification and siting of the generator.
   (a) Explain FOUR considerations in determining the standby generator's specifications and siting.
   (b) Draw a circuit diagram of a basic automatic system which would effect a changeover on mains failure from the mains to a standby generator supplying essential circuits.
   Note: The generator is to be de-activated from supplying the essential circuits two minutes after the mains supply is re-instated. Omit engine controls.

8

The Utilisation of Machines

Complete the following to remind yourself of some important facts that you should remember from the previous chapter.

Three main factors that need considering prior to installing an electric motor are:
1.
2.
3.

Two statutory electrical requirements to comply with the E.A.W. Regs. 1989 are:
1.
2.

Type ________ m.c.b.s are more suitable for motor circuits than types 1 or 2 since their ____________ ______________ characteristics take a ______________ ______of time to operate.

______________ fuses to BS 88 are suitable for backup protection in ________________ circuits, because they are designed to meet motor __________ requirements.

The alignment of two pulleys can be checked using a piece of __________ or a __________ _____________.

Where there is a lot of moisture and dust in the environment, a T.E.F.V (__________ ________ ________ ________) type of motor ____________ would be suitable.

Two items to be inspected on a multiload drive system are:
1. ______________, they may be loose, due to __________
2.______________, they may be worn and have too much ______________.

On completion of this chapter you should be able to:

- calculate the rating of motors for particular applications
- compare the advantages and disadvantages of different motors
- examine the characteristics of different motors for the suitability of particular applications
- compare the starting characteristics and requirements of different motors

It can be seen from the previous chapters that there are many different types of machine. Similarly there are many different applications and loads that require machines to service them. In this chapter situations will be examined to see which machine will meet the described needs. This will be covered in two parts, first to examine how the rating of a machine can be determined; and secondly the other considerations that need to be taken with regard to external factors.

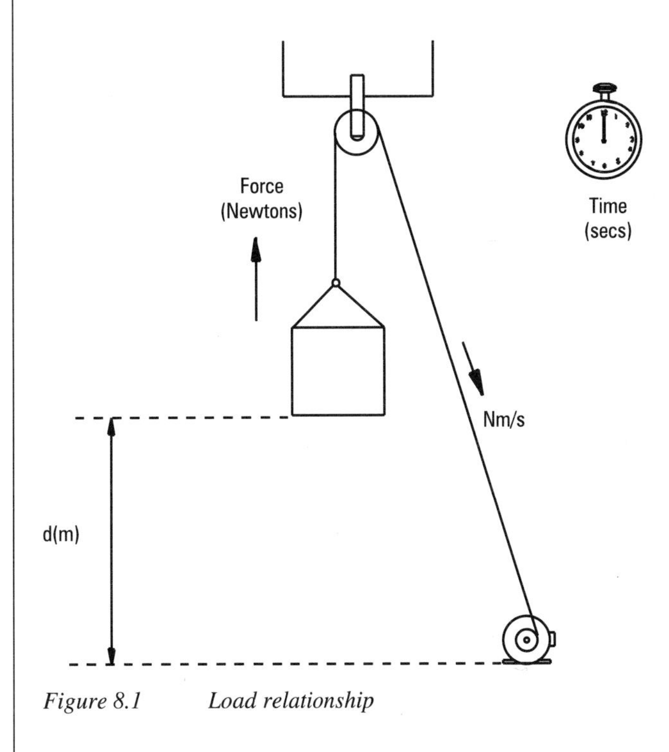

Figure 8.1 Load relationship

Determining the rating of machines

A machine is basically a device that does work. This means that a force is used to move a mass over a distance.

$$W = F \times d$$

Where

W = work done in Newton metres, measured in joules (J)
F = Force to move mass in newtons (N)
d = distance mass is to move in metres (m)

Remember

A force of 9.81 Newtons is required to raise a mass of 1 kg against the effect of gravity.

Example

When a mass of 500 kg is to be lifted through a height of 10 m, the work done is

$$W = F \times d$$

$$= 500 \times 10 \times 9.81$$

$$= 49050 \text{ J}$$

Remember

One watt of power is equivalent to the work being done at the rate of one joule per second (1 W = 1 J/s)

Using the previous example, what power would a motor have to have if the work was to be carried out in 10 seconds?

$$P = \frac{J}{t}$$

$$= \frac{49050}{10}$$

$$= 4905 \text{ W}$$

This assumes that the motor and lifting system is perfect and there are no losses. In fact there will be losses in each part of the system that must be taken into account.

If the above example has losses in the lifting equipment that make them only 85% efficient and the motor is 80% efficient what input power to the motor is now required?

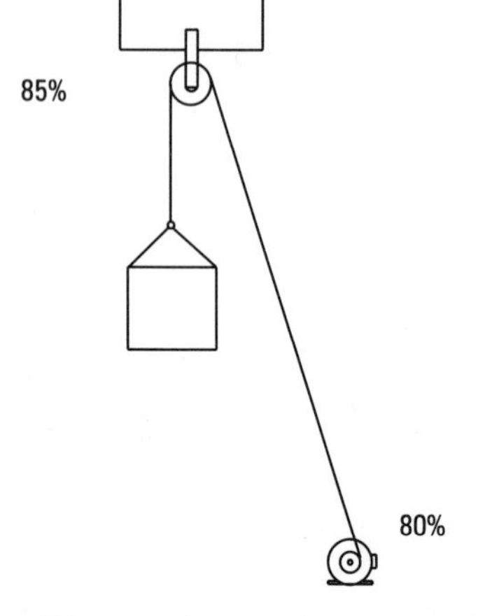

Figure 8.2 Diagram showing the inputs and outputs

The power output of the motor will now require to overcome the losses in the lifting equipment

$$P = \text{output power without losses} \times 85\%$$

$$= \frac{4905 \times 100}{85}$$

$$= 5770 \text{ W or } 5.77 \text{ kW}$$

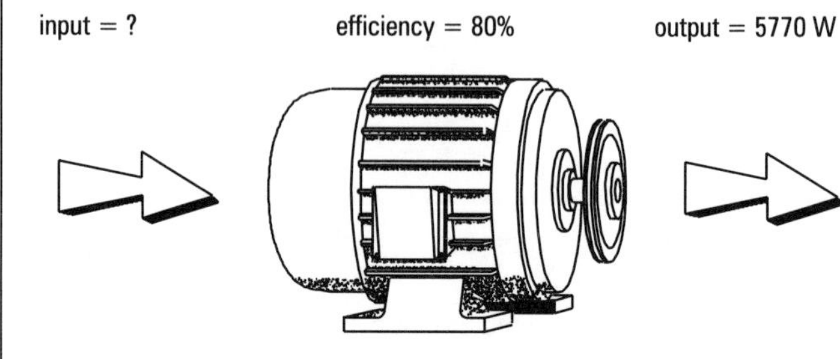

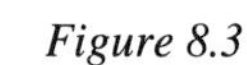

Figure 8.3 Losses in motor

$$\text{Efficiency} = \frac{\text{output}}{\text{input}} \times 100$$

$$\text{Input power to motor} = \frac{5770 \times 100}{80}$$

$$= 7213 \text{ W}$$

This means that the rating of the motor now has to be at least 7213 W instead of 4905 W. There is 2308 W being supplied to overcome the losses.

If this is a d.c. motor, the current rating of the supply circuit can be calculated from this input power.

$$I = \frac{P}{U}$$

If the supply voltage is 240 V d.c. the current rating is

$$I = \frac{7213}{240}$$

$$= 30 \text{ A}$$

If however this is an a.c. motor the power factor that the motor is working at, must be taken into account.

Assume the motor is a single-phase 230 V one working with a power factor of 0.75.

Remember

The power factor should be treated as another form of loss as it increases the current required in the supply cables.

$$P = UI \times \text{power factor}$$

$$I = \frac{P}{U \times \text{power factor}}$$

$$= \frac{7213}{230 \times 0.75}$$

$$= 41.8 \text{ A}$$

When a.c. motors are used, they may of course be 3-phase instead of single-phase. In this case there would be three supply cables instead of two but the current in each would be less. Assuming a three-phase voltage of 400 V and a power factor of 0.8, the current in each supply conductor would be

$$P = \sqrt{3} \times U_L I_L \times \text{power factor}$$

$$\therefore \quad I_L = \frac{P}{\sqrt{3} \times U_L \times \text{power factor}}$$

$$= \frac{7213}{1.73 \times 400 \times 0.8}$$

$$= 13 \text{ A}$$

Try this

A mass of 850 kg is to be lifted vertically to a height of 25 m in 30 seconds. Calculate the line current in the supply conductors if the following motors are considered.

(a) a 200 V d.c. shunt motor with an efficiency of 75%

(b) a 400 V d.c. compound motor with an efficiency of 82%

(c) a 230 V single-phase a.c. motor with a power factor of 0.6 and an efficiency of 78%

(d) a 400 V three-phase motor with a power factor of 0.82 and an efficiency of 79%.

Electric motors are used to drive many different types of equipment and often the rotating movement of the motor shaft is converted into a linear movement to carry the load. The previous example used a linear vertical lift to raise the load to the required height. Other examples can be found in a variety of applications. One is a conveyor, where a motor drives through a pulley system, or gearbox, to create the apparent linear movement of a large flat belt. The conveyor is continuous and does not stop after moving the load a set distance. The load in this case is carried at a rate, i.e. so many metres per second.

Example

A conveyor is designed to carry a mass of 1000 kg at a rate of 0.4 metres/sec. Calculate the work done and the power at 100% efficiency.

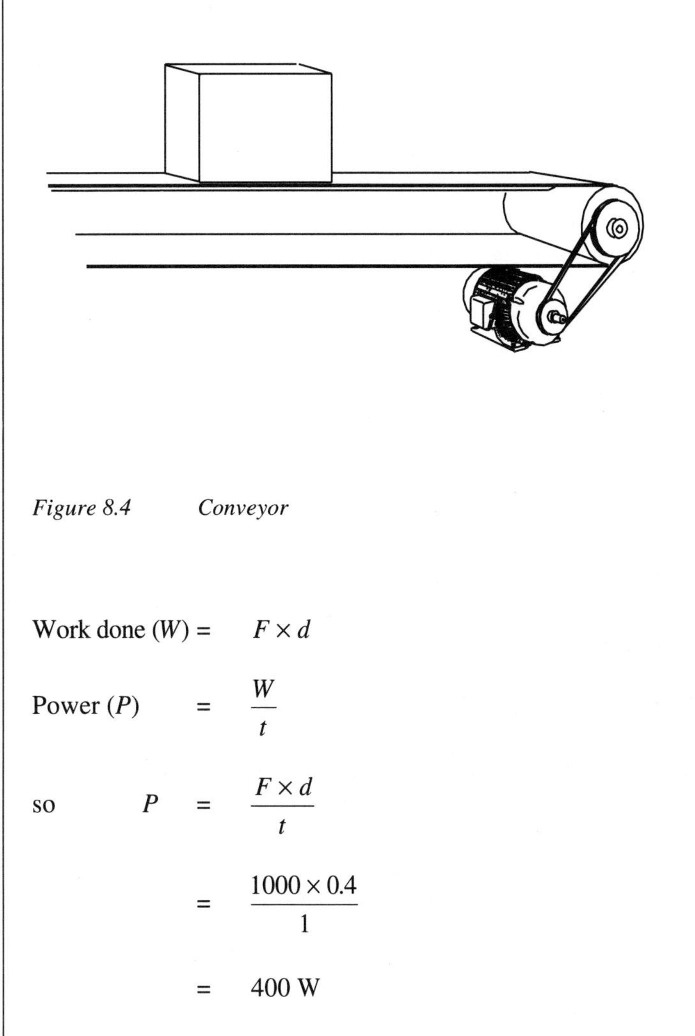

Figure 8.4 *Conveyor*

$$\text{Work done } (W) = F \times d$$

$$\text{Power } (P) = \frac{W}{t}$$

$$\text{so} \quad P = \frac{F \times d}{t}$$

$$= \frac{1000 \times 0.4}{1}$$

$$= 400 \text{ W}$$

Conveyors have losses due to friction of the belt in addition to those in the gearbox and motor. If for this example the conveyor is working on 65% efficiency, the gearbox 80% and the motor 82%, what is the input power to the motor?

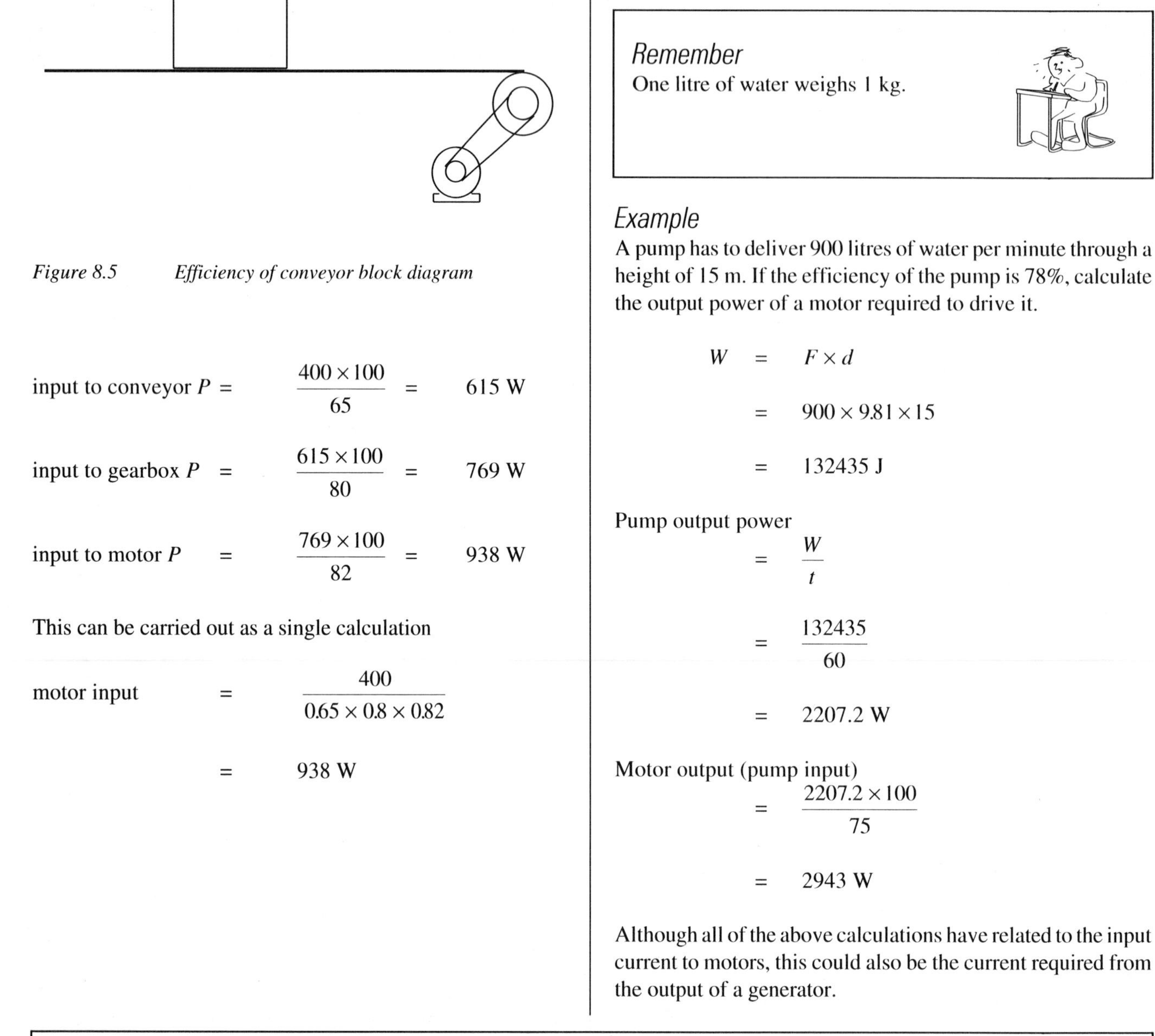

Figure 8.5 *Efficiency of conveyor block diagram*

$$\text{input to conveyor } P = \frac{400 \times 100}{65} = 615 \text{ W}$$

$$\text{input to gearbox } P = \frac{615 \times 100}{80} = 769 \text{ W}$$

$$\text{input to motor } P = \frac{769 \times 100}{82} = 938 \text{ W}$$

This can be carried out as a single calculation

$$\text{motor input} = \frac{400}{0.65 \times 0.8 \times 0.82}$$

$$= 938 \text{ W}$$

Another load that must be considered is the pump where liquid has to be lifted from one height to another.

> *Remember*
> One litre of water weighs 1 kg.

Example

A pump has to deliver 900 litres of water per minute through a height of 15 m. If the efficiency of the pump is 78%, calculate the output power of a motor required to drive it.

$$W = F \times d$$

$$= 900 \times 9.81 \times 15$$

$$= 132435 \text{ J}$$

Pump output power

$$= \frac{W}{t}$$

$$= \frac{132435}{60}$$

$$= 2207.2 \text{ W}$$

Motor output (pump input)

$$= \frac{2207.2 \times 100}{75}$$

$$= 2943 \text{ W}$$

Although all of the above calculations have related to the input current to motors, this could also be the current required from the output of a generator.

Try this

A horizontal industrial conveyor is designed to carry a load of 2.5 tonne at a rate of 0.3 m/s. If the following efficiencies apply to this system – conveyor 70%, drive system 75% and motor 80% – calculate:

(a) the input power to the drive motor
(b) the line current if
  i) a 380 V d.c. shunt motor is used
  ii) a 230 V single-phase motor with a power factor of 0.65
  iii) a three-phase 400 V motor with a power factor of 0.75.

Try this

A pump driven by a 230 V single-phase motor has to raise 400 litres of water per minute through a height of 5 metres. The motor is 78% efficient and has a power factor of 0.65. If the pump is only 60% efficient calculate the current required to supply the motor.

External factors

Throughout this study book many different machines have been considered. Now situations will be studied with a view to recommending which motor would meet the requirements described.

The considerations can be listed as

supplies available

- d.c.
- a.c.
  - single-phase
  - three-phase

load characteristics

- starting currents
  - type of starter
- running
- rapid start/stop

environment

- temperature
- motor enclosure

speed and speed control

- fixed speed
- variable speed

drive system required

- direct drive
- belt

There are often several different types of motor suitable to drive a particular load, but when other factors such as the environment and the drive system required are taken into account the choice may be limited to a single type of machine.

Example

A motor is required to drive a conveyor that is used to take loads of up to 2 tonnes from the back of lorries into a store. The motor will be positioned in the open and may be affected by the weather. The supply is to be taken from either a double-pole 380V d.c. distribution board just inside the store, or a triple-pole and neutral 400/230 V a.c. distribution board on the loading bay. The drive from the motor to the conveyor is by chain, direct to the conveyor drive The conveyor is to run at 0.3 m/s and it can be assumed that the overall efficiency is 60%. It is expected that the conveyor will need to be stopped and started as the loads are fed through. Maintenance of the machine must be considered and kept to a minimum.

Discuss the choice of motor for this situation.

Note: It is important when looking at a situation like this to consider all the facts that are known. There is not necessarily just one motor that meets all the requirements. It is not the actual selection that is important for this exercise, but the process that is used.

Answer

Load considerations

The power that the motor must be capable of supplying is

$$P = \frac{Fd}{t}$$

$$= \frac{2000 \times 0.3}{1}$$

$$= 600 \text{ W}$$

allowing for 60% efficiency of the conveyor

$$P = \frac{600}{0.6}$$

$$= 1000 \text{ W}$$

Assuming each motor will have an efficiency of about 80%

The power input to the motor

$$P = \frac{1000}{0.8}$$

$$= 1250 \text{ W}$$

If a 380 V d.c. motor is used the line current would be

$$I_L = \frac{1250}{380}$$

$$= 3.3 \text{ A}$$

Using a three-phase motor on 400 V a.c. the line current would need to include a power factor of say 0.8

$$I_L = \frac{1250}{\sqrt{3} \times 400 \times 0.8}$$

$$= 2.25 \text{ A}$$

As single-phase 230 V a.c. can be obtained from the triple-pole and neutral distribution board, this must be considered. Again a power factor of at least 0.8 must be taken into account.

$$I_L = \frac{1250}{230 \times 0.8}$$

$$= 6.79 \text{ A}$$

Summary of load considerations

The d.c. supply requires 2 cables designed for 3.3 A, the three-phase a.c. supply uses 3 cables designed for 2.25 A and the a.c. single-phase uses 2 cables designed for 6.79 A.

Starting considerations

As there is no clutch arrangement in use, the motor must be capable of starting on full load. The motor must also be able to keep a constant speed even when the load varies. This means that if a d.c. machine is considered it would have to be either a shunt or compound connected motor. The a.c. three-phase motor would be a cage rotor induction motor and if a single-phase motor was used it would need to be one of the capacitor start type.

There could be problems with the starting arrangements for the d.c. motors as a resistance starter would need to be used to overcome the very high starting currents. Both of the a.c. motors could be started using direct-on-line methods as the currents are not excessive.

Summary of starting considerations

Although two types of d.c. motors could be used, starting requirements for the continuous start/stop could be a problem. Either three-phase or single-phase a.c. motors could be used but the starting current would be less on the three-phase machine.

Environmental considerations

This would affect the type of motor enclosure required and have little or no effect on the motor windings. As the motor is to be in a position exposed to the weather, it would need to be a totally enclosed type. The d.c. motor enclosures would need to have access to the brush gear for maintenance purposes. The mounting of the motor would also need consideration as it has to be suitable for bolting to the frame of the conveyor. It also needs to be adjustable so that any slack can be taken up in the chain drive. The motor shaft must be suitable for fixing a chain sprocket onto for driving the chain.

Summary of environmental considerations

To be suitable for weather requirements the motor enclosure needs to be totally enclosed. Mounting needs to be suitable for fitting to the conveyor frame so that adjustments can be made.

Maintenance considerations

The d.c. motors both have a wound armature and brush gear that need regular maintenance. Also the resistance starters require the contacts checking at regular intervals. Although neither of the a.c. motors under consideration require attention in the same way as the d.c. ones, the direct-on-line starters will require the contacts checking.

Summary of maintenance considerations

As all motor starters will need regular inspection and maintenance the difference between the systems under consideration comes down to the motors themselves. The d.c. machines require regular attention due to the brush gear whereas the a.c. machines do not have the same problem.

Overall summary

The a.c. machines have several advantages over the d.c. and the lower currents required on the three-phase may give this the advantage. There are, however, other considerations that have to be included. Where other motors are used on the same premises it may be possible to match the requirements for the conveyor motor with one of the existing motors. This can have savings in spares and maintenance costs.

Exercises

1. A 1 kW a.c. motor is required to drive a pump in a boiler house. The pump is controlled from a panel and is switched on and off automatically. Discuss the considerations that must be given to whether a single-phase or three-phase motor should be used, assuming both supplies are available. Consideration should include safety aspects for maintenance purposes.

2. A factory has a large number of three-phase loads some of which have a low power factor. It is suggested that a large extractor fan, that is about to be installed, should be fitted with a synchronous motor. Discuss the advantages and disadvantages of this suggestion.

3. The pump shown in Figure 8.6 raises 400 litres of water per minute through a height of 5 metres. If the motor operates at a power factor of 0.7 lagging calculate the current taken from the supply. The motor is 80% efficient and the pump 60%.

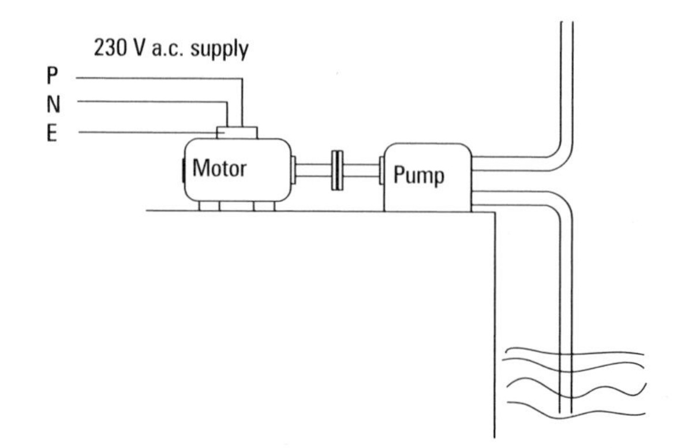

Figure 8.6

4. A pump driven by a three-phase induction motor delivers 9000 litres of water per minute against a total head of 150 m.
If the efficiency of the pump is 75%, the supply voltage is 3300 V, the motor efficiency 92% and power factor is 0.9, calculate the line current.

End Questions

1. (a) Draw the circuit diagram for a long shunt connected compound motor.
   (b) Explain why there is a voltage drop due to the brushes on a d.c. motor.
   (c) A 20 kW 240 V d.c. shunt connected motor has an efficiency of 80%. The shunt field has a resistance of 120 Ω and an armature resistance of 0.15 Ω. If the voltage drop due to the brushes is 2.5 V, determine the
   i) armature current
   ii) generated e.m.f.
   iii) total copper losses

2. (a) Make a labelled sketch of the cross-section of a rotor slot in a double-cage induction motor and explain the action of each cage in producing torque during starting and running.
   (b) Explain why special consideration may have to be given to starting motors in excess of 10 kW.
   (c) Describe how the protection of motors may be provided for
   i) overload circuits
   ii) short circuit currents.

3. It is required to install a large electric motor and its associated control gear on the ground floor of a factory. Although the floor is made of concrete, it is in poor condition and sounds hollow.
   (a) Describe what the procedure is to construct a suitable base for the motor.
   (b) The motor is to drive the load by multiple vee-belts so describe how to align and fix the motor to the new base.

4. A construction site has to have a temporary power supply at 110 V from a generator to provide power for a number of loads.
   The loads are:
   One hoist that lifts a 1 tonne load at 0.2 m/s and works at 60% efficiency, one hoist that can lift 1.5 tonnes at 0.3 m/s and works at 65% efficiency, one set of tungsten halogen flood lights with a total load of 3000 watts.
   (a) Determine the minimum power rating of the generator required.
   (b) Discuss the type of generator that could be used with regard to
   (i) if it should supply a.c. or d.c.
   (ii) the type of prime mover to be used.

5. A diesel powered motor drives a pump that lifts 1000 litres of water per minute through a height of 25 m. The efficiency of the pump is 68%. The existing diesel motor has to be replaced as it is obsolete and it is suggested that an electric motor could be installed in its place. At present the nearest electrical supply is 0.25 km away in a farm building. There is a clear run for overhead cables to be installed over farm land to the position of the pump. Discuss the merits of
   (a) running an electrical supply and installing an electric motor,
   (b) installing a generator and electric motor
   (c) having the diesel motor replaced.

6. (a) State two applications for a pole change motor.
   (b) Explain the basic constructional features of a pole change motor.
   (c) The rotor speed from the nameplate of a pole change cage-rotor a.c. induction motor are given as:
   (i) 950 rev/min
   (ii) 350 rev/min
   For each speed calculate the percentage slip.
   (d) Describe briefly how the torque produced by this motor will be affected when operating at the two different speeds.

Answers

These answers are given for guidance and are not necessarily the only possible solutions.

Chapter 1

p.5 Try this: 17.67 N
p.6 Try this: 2.404 V
p.8 Try this: (1) 244 V; (2) (a) 233.5 V, (b) 32.5 A
p.9 Try this: (1) 304.725 V; (2) (a) 217.6 V, (b) 214.72 V; (c) 210.4 V
p.11 Try this: 600 V
p.15 Try this: (a) 120 W; (b) 460 W; (c) 3193.15 W; (d) 81.67%
p.16 Exercises: (1) 22.5 N; (2) (a) 377.5 V, (b) 100 A; (3) (b) i) 0.52 Ω, (b) ii) 278.2 V; (4) (a) 7739 W, (b) 73.7%, (c) 111.9 Nm

Chapter 2

p.25 Try this: 1250 A
p.30 Exercises: (3) (b) i) 70 A, (ii) 0.214 Ω; (4) (b) i) 143 A, ii) 210.34, V iii) 2.9 kW

Chapter 3

p.35 Try this: 227.5 V
p.37 Try this: 232.4 V
p.38 Try this: (a) 47.5 A; (b) 0.168 Ω
p.40 Exercises: (1) (a) 3 A, (b) 153 A, (c) 156.74 V, (d) 142.5 V; (2) (b) i) 4.09 A, ii) 154.09 A; (3) (a) 255 V, (b) 285 V; (4) (a) 140 A, (b) 2.42 A, (c) 142.42 A, (d) 155.24 V

Chapter 4

p.44 Try this: (1) (a) 0.037 per unit slip, (b) 3.7% slip; (2) (a) 2910 rev/min, (b) 3492 rev/min
p.50 Try this:

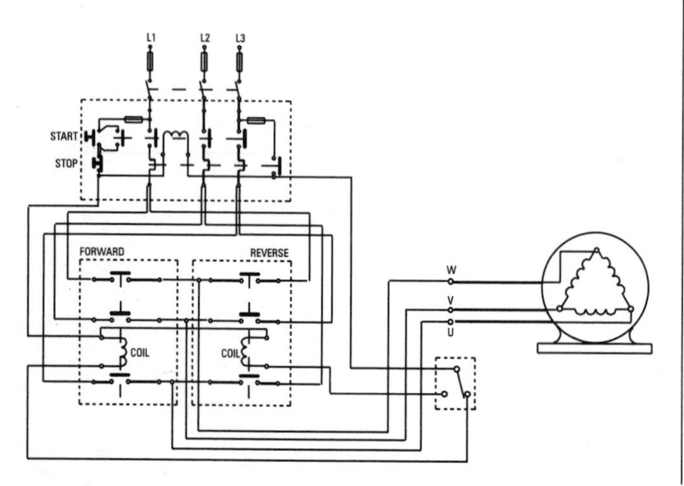

p.62 Try this: (a) 1500 rpm; (b) 1000 rpm; (c) 750 rpm
p.64 Exercises: (2) (b) i) 155.88 kVA, ii) 129.38 kW, iii) 88.88%, iv) 25 rev/sec, v) 8.4% slip

Chapter 5

p.67 Try this: 50 Hz
p.70 Exercises: (1) (b) 750 rpm

Chapter 6

p.76 Exercises: (1) (a) 3000 rpm

Chapter 7

p.79 Try this: (a) 5%; (b) 11.64%; (c) 22.56%; (d) 36%

Chapter 8

p.91 Try this: (a) 46.32 A; (b) 21.18 A; (c) 64.55 A; (d) 15.5 A
p.92 Try this: (a) 1785.7 W; (b) i) 4.69 A, ii) 11.94, A iii) 3.44 A
p.93 Try this: 4.67 A
p.96 Exercises: (3) 4.23 A; (4) 62.19 A

End Questions

p.97 (1) (c) i) 102 A, ii) 222 V, iii) 2.04 kW; (4) (a) 13 kW; (6) (c) i) 5%, ii) 6.7%